METALLURGY OF PURE METALS

METALLURGY OF PURE METALS

Methods of refining pure substances

Jaromír Drapala and Lumír Kuchař

Viva Books

New Delhi | Mumbai | Chennai | Kolkata | Bangalore | Hyderabad | Kochi

First Indian Edition 2009

VIVA BOOKS PRIVATE LIMITED

- 4737/23 Ansari Road, Daryaganj, New Delhi 110 002
  E-mail: vivadelhi@vivagroupindia.net, Tel. 42242200
- 76, Service Industries, Shirvane, Sector 1, Nerul, Navi Mumbai 400 706
  E-mail: vivamumbai@vivagroupindia.net, Tel. 27721273, 27721274
- Jamals Fazal Chambers, 26 Greams Road, Chennai 600 006
  E-mail: vivachennai@vivagroupindia.net, Tel. 28290304, 28294241
- B-103, Jindal Towers, 21/1A/3 Darga Road, Kolkata 700 017
  E-mail: vivakolkata@vivagroupindia.net, Tel. 22836381, 22816713
- 7, Sovereign Park Apartments, 56-58, K. R. Road, Basavanagudi, Bangalore 560 004
  E-mail: vivabangalore@vivagroupindia.net, Tel. 26607409, 26607410
- 101-102 Mughal Marc Apartments, 3-4-637 to 641, Narayanguda, Hyderabad 500 029
  E-mail: vivahyderabad@vivagroupindia.net, Tel. 27564481, 27564482
- Beevi Towers, First Floor, SRM Road, Kaloor, Kochi 682 018, Kerala
  E-mail: vivakochi@vivagroupindia.net, Tel: 0484-2403055, 2403056

www.vivagroupindia.com

Published by arrangement with
Cambridge International Science Publishing
7, Meadow Walk, Great Abington
Cambridge CB1 6AZ, UK

ISBN-10: 81-309-1094-2
ISBN-13: 978-81-309-1094-9

Published by Vinod Vasishtha for Viva Books Private Limited, 4737/23, Ansari Road, Daryaganj, New Delhi-110 002. Printed and bound by Anand Sons, Delhi-110 092.

Contents

Introduction

New areas of advanced science and technology, for example, semiconductors, microelectronics, optoelectronics, superconducting materials, vacuum technology, nuclear metallurgy, space industry and technology, require materials, metals and their special alloys and compounds with high chemical purity and defined physical and structural parameters and specific applied properties. These can be found and achieved mainly in high purity substances and elements with the basic purity of the order of 6N (99.9999%) and higher, i.e., with the total content of impurities and admixtures below 1 ppm (10^{-4}%) and lower.

The metallurgy of pure metals is a subject of university courses concerned with the methods of refining in preparation of high purity substances. Special attention is paid not only to the high degree of chemical purity attainable by chemical or physico-chemical hydrometallurgical methods, such as sorption, extraction, crystallisation from aqueous solutions, electrolysis, and also by pyrometallurgical methods such as crystallisation from melts, evaporation, condensation and transport reactions, electro-transport, diffusion separation of substances, removal of gases from melts and vacuum refining of metals.

In the production of high purity metals by crystallisation refining methods, zone melting and directional crystallisation as the main methods of preparation of defined super purity metals, the controlled redistribution of the impurities and admixtures present in the main substance also takes place at the melt-crystal interface in the single crystal state. The distribution coefficient is the main material parameter which is in fact a measure of the distribution of the admixture in the crystallisation process. The determination of the distribution coefficient, its properties and correlation with the proton number of the admixture of the distribution coefficient is the subject of special attention in this book. The Appendix summarizes the values of the distribution coefficients in binary diagrams, mainly copper, aluminium, iron and its transformation, selected noble, refractory and radioactive metals and lanthanides and semiconductors.

The book is concerned with a wide range of problems in theory and practice of production of high purity metals not only for teaching in the universities of the Czech Republic and Slovakia but also for technical practice and solution of the problems of applied and basic research and examination of the reasons for the formation of primary heterogeneities of castings and their testing.

The authors are grateful to Cambridge International Science Publishing Ltd for publishing the book and to Victor Riecansky for translating and editing the book. This book was written on the basis of the research project of the Ministry of Education, Youth and Sports of the Czech Republic, No. 6198910013 '*Processes of preparation and properties of high-purity and special refined structural materials*' at the Faculty of Metallurgy and Materials Engineering of the VSB–Technical University of Ostrava.

1. PURE METALS AND CLASSIFICATION OF THE METHODS OF PREPARATION OF PURE SUBSTANCES

1.1. Purity, properties and significance of pure substances

A pure substance is a physically and chemically homogeneous substance or a chemical compound consisting of a specific type of atoms or ions or molecules and having only its own typical complex of constant properties. Repeated refining operations are efficient only when they result in the efficient establishment of the properties of the pure substances with the minimum number of operations.

The absolutely pure substance (this term relates to high purity metals and semiconductors and also other materials as regards purity) exists only theoretically. In reality, it is possible to find substances whose purity approaches the absolutely pure substances to different degrees. It may be concluded that the closer the substance is to the absolutely pure condition, the stronger is the effect of its singular properties. When trying to attain absolute purity it is necessary to face insurmountable obstacles. As the number of the impurity atoms in a substance decreases it becomes more difficult to remove them and, consequently, with a decrease of the concentration there are also changes in the potential rate of removal of impurities up to infinity. With an increase in purity there are also problems with maintaining the already achieved purity.

The aim of producing high purity metals is, for example, the effort to determine exact constants characterising the physical properties of these metals. For example, 100 years ago it was possible to produce, as a result of a significant effort, a small amount of high-purity Ag which made it possible to determine more accurately its relative atomic density. This silver was used as an international reference material. In the Twenties of the former century, the method of fractional distillation (rectification) was used to produce pure zinc was the total content of all impurities smaller than 0.05%. The increase of the purity of metals does not influence only the physical properties of metals, for example, density, melting point T_m, boiling point T_b, etc., but also other properties, for example, electrical conductivity, corrosion resistance, and others, which are of considerable importance especially in technical applications of pure metals. In addition, in a number of cases, an increase of the purity of metals results in the appearance of new properties, such as heat resistance, formability, etc., which were unaffected or masked by even very small contents of interstitial and substitutional impurities presented in commercially produced and, consequently, commercial purity metal.

A frequent example mentioned in connection with the actual degree of purity of metals and examination of its effect on the properties is usually represented by aluminium.

The melting point of aluminium T_m^{Al} increases up to a constant value with an increase of the purity of aluminium. For example, aluminium with a content of 99.2–99.5% Al has a melting point in the range 657–658 °C. Purer aluminium with 99.6% Al has a melting point of 658.7 °C, and aluminium with a purity of 99.97% Al 659.8°C. Later measurements taken for pure aluminium with a content of 99.996 % Al [7] gave a melting point of 660.24 °C, i.e. 933.4 K. The latest data [10] indicate a melting point of 660.452 °C for super pure aluminium.

Similarly, the degree of purity of aluminium also determines the density of aluminium: with increasing purity the density of aluminium decreases. For example, a metal with a purity of 99.25% Al has a density of 2.727 kg dm^{-3} at 20 °C, at a purity of 99.4% Al it is 2.706 kg dm^{-3}, and at a purity of 99.75% Al it is 2.703 kg dm^{-3}. The density of pure aluminium with 99.971% Al is 2.6996 kg dm^{-3}, and that of high-purity aluminium with 99.996% Al is 2.6989 kg dm^{-3}.

The recrystallisation temperature of formed aluminium depends strongly on its purity. Aluminium with a purity of 99.99% Al recrystallises at 100 °C, but aluminium with a purity of 99.996% Al shows recrystallisation already at room temperature. Zone-refined super pure aluminium [1,4] with a purity of 99.999% Al after forming at the temperature of liquid nitrogen (–196 °C) starts to show recrystallisation already at a temperature of –50 °C.

An increase in the purity of aluminium also increases electrical conductivity, light reflection, formability and plasticity and also corrosion resistance. This results in new areas of application.

The increase of the importance of high purity substances has been the result of the development of new areas of technology, for example, nuclear technology, aerospace and also microelectronics and optoelectronics, with special techniques of wave transfer and energy quanta and also memory or other properties of pure substances. In these special areas it is not possible to ignore materials defined with respect to purity, not only chemical, but also physical, dislocation-free, structural, nuclear, semiconductor, isotope, etc. Taking into account the areas of application, this type of purity can often be defined more easily as the purity for a purpose.

The purity for a purpose is defined from the viewpoint of the presence of impurities unsuitable or, on the other hand, suitable for obtaining the given or required properties. For example, silicon, used for semiconductor applications, must contain a minimum concentration of admixtures of the first, second, third, fifth and sixth group of the periodic table of elements, and the relatively high concentration of admixtures of the fourth group (Ge) is not harmful, because this element does not have any significant negative effect on the electrical and semiconductor properties of the main element. A nuclear purity material should not contain admixtures unsuitable for nuclear application. As an example, zirconium, used as a cladding material, should not contain hafnium because of its high absorption cross-section with respect to thermal neutrons. The concept of isotopic purity describes the degree of enrichment or separation of isotopes of a specific type, for example, separation of $^{238}_{92}U$ as an unfissionable isotope of natural or enriched uranium from $^{235}_{92}U$ as a fissionable isotope of uranium in processing nuclear fuel. In the case of artificially prepared single crystals, it is necessary to evaluate their structural perfection, the angle of disorientation of grains and the density and distribution of dislocations.

1.1.1. Methods of the description of purity

In most cases, the chemical purity of metals or the concentration of admixtures in the range of pure metals is described by a nine-point system proposed by van Arkel (see the table on page 4).

The designation of the concentration of impurities in ppm: 1 ppm = 10^{-4}% of admixtures, i.e. the purity at which for 1 million (10^6) atoms of the main substance there is 1 atom of the admixture. For even lower concentrations, it is necessary to use the term ppb, i.e. parts per billion; 1 ppb = 10^{-7}% of admixtures, i.e. for 1 billion (10^9) of the atoms of the main substance there is 1 atom of the admixture. For super low concentrations of the admixtures, it is recommended to use the unit ppt, i.e. parts per trillion, 1 ppt = 10^{-10}% of admixtures, i.e. for 1 trillion (10^{12}) of the atoms of the main substance there is 1 atom of the admixture.

This extreme dilution is used in practice in the identification of

Designation	Purity [%]	Impurity conten [%]	
1 N	90 – 99	10 – 1	
2 N	99	1	
3 N	99.9	0.1	1000 ppm
4 N	99.99	0.01	100 ppm
5 N	99.999	0.001	10 ppm
6 N	99.9999	0.0001	1 ppm
7 N	99.99999	0.00001	100 ppb
8 N	99.999999	0.000001	10 ppb
9 N	99.9999999	0.0000001	1 ppb
10 N	99.99999999	0.00000001	100 ppt
11 N	99.999999999	0.000000001	10 ppt
12 N	99.9999999999	0.0000000001	1 ppt

trace content of isotopes of radionuclides in the living environment. In this case, the unit ppt is used at present for the measurement of exhalation of the radon isotope from soil (permission for building). For example, it is also used for accidental escape of substance from the fuel cycle a nuclear power station or contaminated gaseous products or fission fragments.

Another method of definition of purity, in particular in semiconductor technology, is the unit cm^{-3} which expresses the total number of the atoms of impurities and admixtures in 1 cm^3 of the main semiconductor. The current requirements on the silicon suitable for microelectronics permit the maximum total content of 10^{14} of all atoms of the admixtures in 1 cm^{-3} of silicon, which represents the purity of approximately 8 N, i.e., approximately 10 ppb of the present impurities. Silicon with a diamond-like structure has a lattice constant of $a = 5.429 \cdot 10^{-10}$ m, which represents a total of $5 \cdot 10^{22}$ of silicon atoms in 1 cm^3. After doping with 1 ppm of a dopant (microalloying addition), $5 \cdot 10^{16}$ atoms of the admixture penetrate into Si. In this case, only 1 atom of the dopant is available for 1 million atoms of Si. For the given purity of semiconductor Si of 8 N only 'one' atom of the harmful impurity is permissible for 'one hundred million' of Si atoms.

1.1.2. Effect of impurities on the properties of substances

Nuclear engineering requires functionally essential pure metals, starting with uranium as the main and, at present, already classic nuclear fuel material, through coating materials, moderators, shielding materials up to structural materials and alloys for the fabrication of different parts of nuclear reactors. These materials must not or, on the other hand, must absorb thermal neutrons which maintain the course of the controlling fission reaction of $^{235}_{92}U$ which, in addition to the energy gain, produces 2 or 3 neutrons required for the further fission process. Therefore, nuclear fuel in particular should not contain admixtures which would absorb neutrons.

The suitability of different chemical elements for absorbing thermal neutrons is usually characterised by the effective cross-section for the absorption of thermal neutrons σ_a [m^2] which basically includes part of the term nuclear purity. In this case, it is important to consider the so-called isotope purity because generally efficient cross-sections for the absorption of thermal neutrons of the individual isotopes of the same element mutually differ, sometimes by several orders of magnitude. For example, the value of σ_a for hydrogen isotopes $^1_1H = 0.33$ and $^2_1H = 0.00055 \cdot 10^{-28}$ m^2, or in the case of tungsten isotopes $^{190}_{74}W = 10$, $^{182}_{74}W = 0.5$, $^{184}_{74}W = 0.0024$ and for $^{186}_{74}W = 37 \cdot 10^{-28}$ m^2.

A typical example of nuclear purity is the application of zirconium, characterised by a small absorption efficient cross-section $\sigma_a^{Zr} = 0.18 \cdot 10^{-28}$ m^2, as a coating material of fuel elements. In the natural condition, zirconium contains 2–3% of hafnium which is its chemical analogue and always accompanies zirconium in the nature. The absorption efficient cross-sections of zirconium and hafnium mutually differ by the ratio $\sigma_a^{Zr} : \sigma_a^{Hf} \sim 1:600$. This means that the almost trace content of hafnium in zirconium makes the latter unsuitable for application in the natural condition in nuclear technology. Therefore, nuclear technology has necessitated the application of zirconium greatly depleted in hafnium. Of course, this results in a need for the development of special methods of refining of zirconium, i.e. the removal of hafnium from zirconium. However, this has resulted in a number of quite serious difficulties because the chemical properties of zirconium and hafnium are very similar.

Another suitable example of nuclear purity is a structural material of the primary circuits of nuclear reactors made of stainless steel containing nickel in which cobalt is usually present as a genetic accompanying admixture. If stainless steel is used in chemical industry, the presence of Co is not detrimental. Co is characterised by a relatively large absorption efficient cross-section $\sigma_a^{Co} = 35 \cdot 10^{-28}$ m^2 and the neutrons, absorbed by cobalt, lead to the formation of the radioactive isotope $^{60}_{27}Co$ with the decay half-time of 5.2 years, which is characterised by dangerous gamma radiation. Therefore, cobalt is undesirable with respect of the absorption of neutrons, and also owing to the fact that steels, subjected to long-term neutron radiation in thermal reactors, are also a source of high-intensity radiation of $^{60}_{27}Co$ which from the viewpoint of operation complicates simple maintenance and, also the repair of the reactor. Therefore, in the primary production of nickel it is necessary to include efficient, multistage refining to decrease the cobalt content as the undesirable admixture in nickel in stainless steels used in reactors. Similarly, the presence of copper is also an undesirable factor. Under the effect of the neutron flux copper greatly decreases the mechanical properties of structural components of the reactor.

In nuclear power engineering, in particular, in fast reactors, it has been necessary to use metallic melts for the transfer of heat. Liquid metals Na (t_m^{Na} = 97.8 °C) or Li (t_m^{Li} = 180 °C) or certain eutectic Na–K alloys (with the eutectic temperature of approximately 11 °C) are characterised by low vapour tension at elevated temperatures and are used for the transfer of heat from the heated sections of the nuclear reactor. For these applications, the metals must be purified to remove interstitial and metallic admixtures, i.e. they must be refined to a high degree of purity in order to prevent, in particular, pitting corrosion in steel pipes and exchangers.

In reactor and space technology, extensive use is made of important heat-resisting, high-melting and, at the same time, high-toughness and formability metals and pseudoalloys of metals such as titanium, very light but also toxic beryllium, and other metals. Satisfactory formability of these metals, which is an essential condition for forming these metals, can be achieved only after deep refining to remove admixture elements resulting in brittleness of the metals, such as, for example, interstitial elements oxygen, nitrogen, hydrogen, and carbon. Therefore, operation with beryllium is carried out in protective suits in an inert medium.

Applications in microelectronics include the use of many types of high-purity and structurally defined materials, including elementary semiconductors or intermetallic phases and chemical compounds. The requirements on the refining of semiconductor materials to remove admixtures are extremely high. The semiconductor materials include a number of the highest purity compounds, primarily silicon and germanium, and also the elements forming semiconductor compounds of the type $A^{III}B^{V}$ such as AlSb, GaAs, GaP, InP, type $A^{II}B^{VI}$ such as sulphides, oxides, selenides, tellurides CdTe, CdS and $A^{I}B^{VII}$ and many more complicated (Bi_2Sb_3) and ternary ($Cd_xHg_{1-x}Te$) or polycomponent compounds of various elements. In the technology of preparation of microelectronic semiconductor components special attention is given, in addition to microalloying, to the diffusion processes to ensure the formation of *p–n* junctions, and also to the epitaxy and evaporation of superfine chemically and structurally defined nanolayers. These technologies require extremely high purity of the initial compounds and chemical substances.

Semiconductor materials have replaced in electronics vacuum valves used as rectifiers, have enabled the direct conversion of thermal and light energy, for example, the conversion of solar to electric energy. In addition, semiconductors have been used to solve a number of problems in the area of automation and remote control, they are playing the functions of memory components in computing technologies, etc. This list must be supplemented by the area of miniaturisation in microelectronics and other special branches. In order to evaluate the important role played by the semiconductors at the present time and in future technologies, it is necessary to examine the problems as to why it is necessary to carry out the efficient refining of semiconductor materials to remove impurities to the highest possible and obtainable degree of purity.

Conductivity is divided into hole and electron conductivity. However, semiconductor crystals often genetically contain in many cases impurities and admixtures whose presence, already in the trace amount, has a significant negative effect on the change of the electrical properties of the semiconductor. Depending on the nature of the atoms of the admixture, the semiconductor may be characterised by the formation of a surplus of electrons or by a shortage of electrons, i.e. the formation of holes.

The intensity of the effect of the admixtures on the electrical conductivity of semiconductors may be indicated by the fact that 1% of the atoms of the admixture may increase the conductivity of the semiconductor at room temperature by up to 1 million times. Therefore, when refining the semiconductor compounds to remove the admixtures, it is necessary to eliminate random admixtures to enable in subsequent metallurgical treatment intentional alloying of the compounds even with a very small or defined content of the admixture elements of the *p*- or *n*-type. This microalloying is carried out to ensure the required semiconducting properties of the compounds. The high degree of purity must be characteristic not only of the actual semiconductors, such as germanium and silicon, but also of a large number of metallic elements and non-metals used either for the microalloying of germanium or silicon, the formation of superfine layers by epitaxy or evaporation, or for the synthesis of intermetallic compounds with the semiconducting properties, and also for compounds using semiconductor technology such as contact materials, solders and conducting parts or joints.

In addition, it is also important to describe the area of electronic devices and equipment operating in the conditions of deep vacuum (10^{-3} to 10^{-8} Pa). In addition to the general high purity of the materials used for the fabrication of this type of equipment, it is also necessary to ensure the minimum gas content of electrically conducting and high-melting metals, such as for example, Cu, Ni, Co, Ag, W, Mo, Ta, etc. The extremely low gas content is important not only for ensuring high vacuum in equipment (for example, i.e. on ion pumps coated with a gold layer) but also for ensuring their long life and long-term reliability (for example, in space technology).

2. GENERAL CHARACTERISTICS AND CLASSIFICATION OF METHODS OF SEPARATION AND REFINING OF SUBSTANCES

In order to obtain high-purity metals and semiconductor compounds, it is necessary to carry out deep refining of the main component to remove admixtures. Therefore, it is essential to use the efficient combination of the methods based on different physical, chemical and physical–chemical properties of substances, ensuring the separation of impurities and admixtures from the main compound. In particular, the problem of refining and purification is very complicated in cases in which the refined material and the admixture are characterised by very similar physical–chemical characteristics, and when the behaviour of the compound of these metals in a specific chemical or physical process is similar. These problems must be examined in detail when it is necessary to carry out separation or purification of various couples of metals, such as zirconium and hafnium, niobium and tantalum, tungsten and molybdenum and, in particular, in the separation of rare earth metals, spent nuclear fuel, etc.

The methods used in current technology used for the preparation of high-purity materials are many and are often based on a small difference of the physical–chemical properties and behaviour of both the actual elements and compounds of these elements. For the separation of the admixtures and impurities from the metals and semiconductors, it is necessary to use hydrometallurgical and also pyrometallurgical and electrometallurgical methods and techniques based on:

- the different solubility of the individual elements or their compounds (liquid extraction, crystallisation from the solution, directional and zone crystallisation from melts, selective precipitation, dissolution of gases in metals);
- the different volatility of elements and their compounds (distillation, sublimation, rectification, iodide refining, vacuum refining);
- the difference of ionisation potentials (distillation in the form of sub-compounds);
- electrochemical distribution potentials (electrolysis, electrolytic refining, amalgamation electrolysis);

- sorption characteristics (adsorption, ion exchange, chromatography);
- the difference of the oxidation and reduction potentials (selective oxidation and reduction);
- diffusion rate (separation of the isotopes on cascades);
- electromagnetic properties (separation in a magnetic field).

A concentration gradient of the admixtures in the adjacent phases always appears:

- during extraction in two liquid phases;
- during solidification the difference in the composition of the liquid and solid phases at the crystal–melt interface;
- during distillation the difference in the composition of the liquid and gas phases.

Consequently, there is a large number of refining processes which must be combined when selecting the most suitable procedures for each specific basic substance.

2.1. Main methods of production of pure metals and substances

1. Evaporation and condensation	
Separation of highly volatile substances and impuritities	Evaporation Distillation Sublimation
Separation of evaporated main component	Fractional distillation or condensation
Separation of mixtures by re-distillation	Rectification Transport reactions
2. Separation between two phases	
Separation between solid and liquid phase of main component	Zone melting Directional crystallisation Freezing-out
Separation between two solvents	Liquid–liquid extraction
3. Separation on the basis of different solubility	
	Fractionated crystallisation
4. Exchange reactions	
	Ion exchange Chromatography
5. Chemical reactions	
Separation of non-noble components by oxidation	Partial chlorination refining
Reduction by gases	Reduction by hydrogen, CO, CH_4, etc.

	Deoxidation
Electrochemical reduction	Electrolysis of aqueous solutions
	Amalgamation electrolysis
	Salt melt electrolysis
Reduction by metals	Metallothermy
Thermal dissociation	Dissociation of iodides

Below, we propose the classification of the main processes of refining metals and certain substances to remove impurities and admixtures [7]. In this classification of the refining procedures for the separation and refining of substances, the processes are divided into groups 1 to 8 on the basis of the physical–chemical properties utilised in the separation of components.

1. Sorption

- adsorption
- ion exchange
- chromatography

2. Extraction

- liquid extraction

3. Crystallisation

- crystallisation from solutions
- crystallisation from melts
- directional crystallisation*
- zone melting*
- freezing-out

4. Evaporation and condensation

- sublimation
- distillation
- rectification*
- distillation by means of transport chemical reactions*

5. Electrolysis

- electrolytic precipitation with preliminary cleaning of the electrolyte*
- electrolytic refining*
- amalgamation electrolysis*
- electric transfer*

6. Diffusion

- diffusion
- thermal diffusion
- high-voltage dialysis

7. Selective precipitation, oxidation and reduction

- selective precipitation
- selective oxidation and reduction

8. Removal of gases

- vacuum extraction
- electron beam melting*

Comment: the processes indicated by * are used for higher degrees of refining.

This classification is simple and natural. However, a significant disadvantage is that one typical group may contain processes completely differing in their nature. For example, the group of the distillation processes, in which the components are separated on the basis of their different volatility, include both simple distillation and sublimation processes and also more complicated processes of the type of rectification or transport reactions.

The individual refining processes may be used in appropriate combinations either directly for elements and also for chemical compounds of these elements. It should be mentioned that if a simple substance is characterised by a high boiling point and efficient distillation of this substance would be difficult, this substance can be transformed into a component which can be distilled far more efficiently. For example, the melting point of Ga is 29.8 °C but Ga is characterised by a very low vapour tension to temperatures of ~1500 °C. This is greatly inconvenient for the distillation operations, but Ga can be transformed to, for example, trichloride $GaCl_3$, with a boiling point of 201.3 °C, and purification of this component by the distillation method can be carried out quite easily and using a simple procedure: from the refined chloride, the metal can be precipitated by electrolysis from the aqueous solution or reduced by hydrogen. In all cases in which it is not possible to purify the simple substance or compounds as such, they can be defined by transformation to an intermediate component. This intermediate component is subsequently efficiently purified and then transformed back to the initial substance (for example, the reaction utilising subchlorides of refining metals in transport reactions).

In order to produce pure metals and semiconductor materials, purification is usually carried out in two stages:

1. Deep purification of chemical intermediate compounds resulting in the formation of pure metals and semiconductors;
2. Additional refining of the resultant metals and substances.

The first stage of purification of the substances is based mainly on the processes of hydrometallurgy; electrometallurgy and distillation:

1. Absorption and ion exchange

2. Extraction with organic solvents, liquid extraction
3. Re-extraction
4. Multiple precipitation of elements in the form of low-solubility compounds
5. Precipitation of admixtures from solutions
6. Multiple recrystallisation from solutions
7. Electrolytic refining from aqueous or organic electrolytes
8. Rectification

The products of the first stage of purification are usually chemical compounds, cathodic metals, powder metals and distilled compounds with a commercial purity of up to 3 N.

The second stage of purification is characterised in most cases by a combination of pyrometallurgical and electrometallurgical procedures:

1. Electrolytic refining of aqueous or organic electrolytes or salt melts;
2. Distillation and rectification
3. Distillation through sub-compounds and subsequent reduction
4. Thermal dissociation of unstable compounds
5. Directional crystallisation
6. Zone melting
7. Electron beam melting in high vacuum for separation of volatile admixtures and gases, plasma melting, ion melting in inert or reaction inert atmospheres, etc.

In some cases, it is possible to obtain approximately the same result in the refining metals using different purification methods. For example, to obtain aluminium with a purity of 99.999 to 99.9999 % Al for transitional contacts of silicon power rectifiers, one can use either the zone refining of electrolytically refined metal or distillation in the form of a sub-component, or additional electrolytic refining in inorganic solvents.

An important aspect is the selection of the most efficient purification method on the basis of the required properties of the produced material and the productivity of methods for each specific refining case.

However, in most cases it is not possible to obtain the required results in purification of metals and semiconductors using only one method in each purification stage and, consequently, it is necessary to combine efficiently the individual refining methods. This is especially the case in the mutual separation or refining of rare earth metals (lanthanides) where, because of the differences in the required properties and defined purity, it is necessary to combine efficiently the hydrometallurgical extraction and ion exchange methods with the methods of fractional crystallisation and methods of electrolytic refining in salt melts, pyrometallurgical methods of selective oxidation and metal-thermic reduction should be combined with methods of selective evaporation and condensation with the methods of zone refining and preparation of crystals of the individual lanthanides with a defined degree of purity.

The rules which must be maintained in the processes of preparation of pure metals:

1. The individual stages of the refining processes must be selected in such a manner as to

a) ensure that it would be possible distinguish continuously during refining processes the chemical, physical–chemical and physical properties of refined substances;

b) ensure that the number of the individual basic operations, resulting in the required purity of the substances, is as small as possible for the given combination;

2. The processes must be easy to realise from the technical viewpoint:

a) if resistant structural materials are not available, it is necessary to find a solution preventing contact between the walls of equipment and refined metal at higher temperatures;

b) it is important to prevent the penetration of impurities (contamination) from the surrounding environment and from the atmosphere.

3. Purification must be carried out as long as possible by the chemical method

a) it is necessary to prefer processes with a low consumption of chemical agents

b) it must be possible to produce chemical agents with high purity;

c) it is important to ensure efficient capture of waste.

4. The refining of substances to remove the most powerful impurities must be started as soon as possible, but not too early because the attained purity should be maintained

5. The processes of production must be characterised by high productivity, economic efficiency and safety

6. It is necessary to ensure a continuous working procedure with continuous production.

Table 1 Approximate purity of metals obtained in selected refining methods

Method	Metal	Total content of impurities [ppm]	Method	Metal	Total content of impurities [ppm]
Oxidation with additions	Mg	10	Dissociation of iodides	Zr	100
	Mn	10		Hf	150
	Pt	30		Cr	200
	Au	20–30		Th	100
				Al	30
Fractional distillation	Cd	10	Electrolysis	Cu	10
	Mg	5		Fe	100
	Zn	10		Ga	0.1
	Ca	10	Zone melting	Si	0.01
	K	10–30		Ge	0.01
	Li	10	Vacuum melting	Cu	10
	Na	10		Fe	10
				Ni	10

3. SORPTION

3.1. Significance of sorption processes in the separation and refining of substances

Sorption is a process characterised by the absorption of gases, vapours or dissolved substances by the sorption agent, i.e. the sorbent, at the liquid–gas, solid–gas or solid–liquid interface. Sorption includes both adsorption, characterised by bonding on the surface of the solid or at the phase boundary, and absorption, characterised by absorption by the entire volume of the sorbent. The processes represent the basis of the methods of adsorption, ion exchange and chromatography [11]. Sorption is used extensively in the separation of rare earth metals–ion exchange, separation of zirconium and hafnium, extraction of uranium from highly diluted solutions of uranium compounds, the preparation of demineralised (deionised) water for semiconductor technology, selective separation of the individual admixtures from the solutions of the main substance, etc.

3.2. Adsorption

Adsorption is a process taking place on the surface of the solid substance or liquid (adsorbent) based on the increase of the surface concentration of the molecules of the adsorbed substance (adsorbate) which may be represented by a gas or a substance dissolved in the solution. The adsorption of the molecules of the adsorbate on the adsorbent is caused by binding forces between the particles of the surface of the adsorbent and the molecules of the adsorbate falling on the surface. The bonded molecule remains on the surface for a specific period of time, referred to as the dwell time, which depends on the bonding energy and temperature. The molecules, adsorbed on the surface, result in the formation of a gas layer with special properties: the adsorbed molecules during the dwell time carry out random thermal motion in relation to the plane of the surface. On the surface of the adsorbent there are absorbed substances decreasing its surface tension in relation to the surrounding medium. To obtain a strong adsorption effect, it is necessary to ensure the maximum possible surface of the adsorbent; the materials with a highly ragged surface, porous and spongy substances are suitable for this purpose.

The most frequently used adsorbents in practice include specially processed active coal (charcoal or animal charcoal) with high porosity and, consequently, the extremely large surface of the pores. For example, the internal surface of pores in 1 g of efficiently absorbing active coal reaches 400–900 m^2. The nature of porosity is also important. When producing pure compounds of metals, semiconductor materials and some other substances, good results are obtained with highly porous or highly dispersed silica gel, clay, kaolin, and some other alumosilicates.

If the bonding absorption forces are relatively low, the adsorbed molecules and, consequently, surface particles, retain their individual properties. This is physical absorption. When, during absorption, a molecule receives or transfers an electron or dissociates into an atom, which is chemical bonded with the surface particles, chemisorption takes place. In chemisorption, bonding is far stronger and the appropriate values of the absorption heat are relatively high. Chemisorption has the nature of a chemical reaction and the absorption heat corresponds to the reaction heat of this reaction.

The amount of the dissolved substance, adsorbed by a certain amount of the given adsorbent, depends on the nature of the dissolved substance and, consequently, the conditions of the process, i.e. the concentration of the dissolved substance and temperature. The dependence of the surface concentration of the molecules in absorption on the volume concentration of the molecules above the surface of the adsorbent in the equilibrium condition is described by absorption isotherms (Fig. 1). The maximal amount of the substance adsorbed by the given amount of the adsorbent at the moment of its saturation (region III in Fig. 1) characterises the absorption capacity of the adsorbent. In

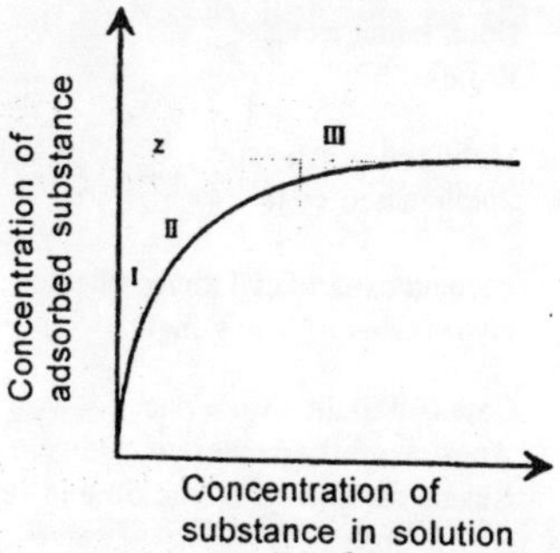

Fig. 1 Adsorption isotherm.

absorption from the solutions, the substances characterised by a lower solubility in the given solvent are usually characterised by stronger absorption.

Absorption is utilised in analytical methods, in separation of mixtures, separation of dissolved compounds from solutions, in vacuum technology, and in catalytic reactions.

3.3. Ion exchange, ion exchangers

The methods of ion exchange are based on the reversible exchange of ions between the external liquid phase and the solid ion phase. The solid phase consists of an insoluble but permeable polymer network or crystal lattice, containing bonded groups with a charge and mobile counter-ions with the opposite charge. These counter-ions may be substituted by other ions from the external liquid phase. If selective exchange forces operate, enrichment with one or several components takes place. The method is suitable for substances which are at least partially ionised [12].

Ion exchangers (ionex, ionite) are macromolecular, network compounds with a suitable grain size, containing a large number of ion groups as substitutes on the framework of the polymer, capable of exchanging counter-ions.

According to the origin, there are the following types of ion exchangers (Table 2):

Table 2 Commercial ion exchangers

Origin	Type of skeleton	Active group	Example, commercial name
Natural	In	C	Zeolites, glauconites
		A	Dolomites, apatites
	Org	C	Coal, humic acids
		A	Protein
Modified	In	C	Stabilised green sand
Natural	Org	C	Sulphonated coal
	In	C	Permutites–artificial alumosilicates
		A	Hydroxides of heavy metals
Artificial		C	Catex–Wofatit, Amberlite, Dowex, Dualite, Ostion, Zeolite
prepared		A	Anex–Wofatit, Amberlite, Dowex, Varion, Ostion
resins		R	Resins capable of oxidation and reduction

In – inorganic, Org – organic, C – catex, A – anex, R – redox.

Natural – compounds present in nature, mineral ion exchangers, humic acids.

Modified natural substances – stabilised green sand, coal–cellulose.

Artificially prepared substances – artificial alumosilicates, organic and resin ion exchangers.

On the basis of the origin and properties of the functional groups, the ion exchangers can be divided into:

Cation exchangers – cation exchange resins, containing acid functional groups;

Anion exchangers – anion exchange resins, containing basic functional groups;

Amphoteric ion exchangers–mixed, containing both types of functional groups on the same macromolecular framework

Redox resins, redox ion exchangers – conventional ion exchangers cap ions capable of oxidation and reduction, redox ion exchangers-resins containing the group capable of reduction and oxidation bonded directly on the polymer framework of the resin

Selective ion exchangers – containing active groups capable of reacting only with a small group of the ions, in the ideal case only with one type.

According to the external form:

granular – spherical particles, pearl-shaped particles, or networks of an irregular shape, the size of the particles from several millimetres to 10^{-2} or 10^{-3} mm;

non-granular – ion exchangers in the form of membranes, filters, papers, fibres, foam, sheets, cloth, etc.

The artificially produced ion exchangers based on synthetic resins produced by polycondensation or polymerisation are used on an increasing scale because of their excellent properties, for example, high exchange capacity, chemical resistance and mechanical strength. Their structure is similar to plastics and electrochemical activity is the result of the introduction of active groups. The skeleton is formed by open carbon or cyclic chains. Modern ion exchangers contain as the matrix, for example, polystyrene with a certain amount of divinyl benzene. The most important exchange active groups in organic ion exchangers are:

cation exchangers: active group: sulphonic ($-SO_3H$), carbonyl ($-COOH$), phosphonic ($-PO/OH/_2$), hydroxyl ($-OH$), arsonic ($-AsO/OH/_2$)

anion exchangers: active group primary amine ($-NH_2$), secondary amine ($-NH$); ($-NH^+/CH_3/_3$) group

amphoteric exchangers: active group ($-N/CH_2A-COOH/_2$), (HO_3S – matrix $-N^+/CH_3/_3$)

The reaction

Cation exchangers:	$RH + Na^+ \rightleftarrows RNa + H^+$	(1)
Anion exchangers:	$2\,RCl + SO_4^{2-} \rightleftarrows R_2SO_4 + 2Cl^-$	(2)
Amphoteric ion exchangers:	$HR_2OH + KCl \rightleftarrows KR_2Cl + H_2O$	(3)

where R is the carrier – the skeleton of the exchanger with ions. The ion exchange between the solution and the exchanger is reversible.

The ion exchange process takes place in the majority of cases in a vertical column represented by, for example, a glass tube of suitable dimensions filled with an ion exchanger, with the supply of a liquid medium on the surface of the column of the exchanger, Fig. 2.

The dynamic method of application of the ion exchangers consists of three stages:

- sorption in the column – the exchange of the ions between the solutions supplied into the column and the ions originally bonded by the exchangers with a possible addition of a complex making agent to ensure selective sorption;
- rinsing the column of the exchanger in order to complete sorption with water or another solvent in order to rinse-out the remainder of the original solution;
- desorption – elution, i.e. reversed displacement of the ion, bonded on the ion exchanger, by means of a suitable elution solution. During elution, the eluent with the content of one or several types of ions adsorbed in the column in the first stage of the exchange cycle flows out of the column. The suitable composition of the dilution solution can also be obtained by the so-called selective elution, i.e. the case in which only one type of ions is released from the mixture of the ions sorbed by the ion exchangers. As a result of elution, the ion exchanger is returned back to the working condition.

In the ion exchange, one equivalent of a different ion with the same type of charge is released by the ion exchanger into the solution per every equivalent of the specific ion, and adsorbed from the solution.

Physical-chemical characteristics of ion exchangers

Exchange capacity is the content of counter-ions or the capacity to retain the counter-ions in the unit volume or unit mass of the resin. The total capacity is the theoretical value for a specific material and can be calculated from the number of functional groups in the mass unit of the dry substance.

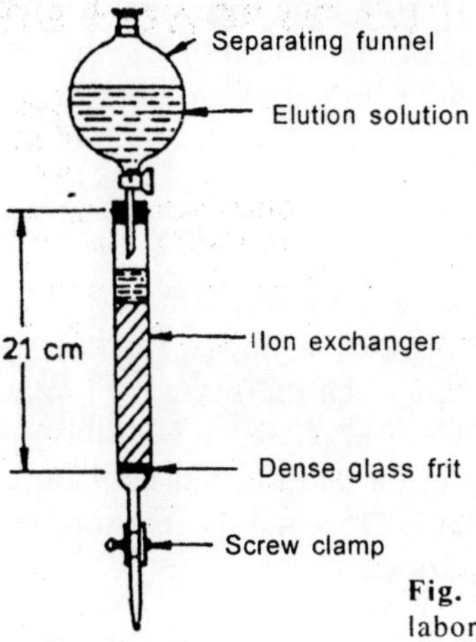

Fig. 2. Ion exchange process, typical laboratory ion exchange column.

Swelling capacity – the resins are different gels which in the dry condition efficiently adsorb water and other polar solvents in which they are immersed. During absorption of the liquid, the gel structure expands or swells while the stresses of the expanded polymer network are not the same as the osmotic pressure. The counter-ions and ionised groups inside the resin determine the extent of hydration of the ion exchanger and the structural skeleton must ensure the stability and durability of the resin in practical application in water, acid and basic solutions and organic solvents. After swelling, the ion exchanger is ready for use.

Acidity, basicity – the chemical properties of the ion exchanger are basically determined by the type and amount of free and strong ions and by the structure of the macromolecular network, in particular, the degree of cross-linking. The type of active, ionogenous groups depends on the strength of the acid or basic properties of the ion exchanger. From this viewpoint, cation exchange resins are divided into strong or slightly acid and anion exchange resins to strong or slightly basic.

Chemical stability – the resistance to the effect of oxidation and thermal effects of solutions of alkali or acids, hydrogen peroxide, etc.

Selectivity – every ion exchanger is characterised usually by a specific degree of preferential sorption for a specific type of ions. Selectivity may be defined as the differentiating capacity to bond different ions by different forces. The selectivity of the ion exchanger is affected by the structure of the resin, the concentration of univalent ions and the valency of another ion. During charging operations, including the exchange of the counter-ions of the resin in the hydrogen form of the ions M with charge n from the solution of an electrolyte, the following reaction takes place:

$$H^+R^-_{/S/} + \frac{1}{n}M^n_{/N/} \rightleftarrows \frac{1}{n}M^{n+}R^-_{n/S/} + H^+_{/N/} \tag{4}$$

where R are the negatively charged exchange groups in the resin network, the argument /N/ indicates the solution, the argument /S/ the solid phase. Equilibrium is established when there are no longer statistical changes in the ratio M^{n+}/H^+ in the resin phase. In equilibrium, the concentrations of the ions in the solution and the ion exchanger are determined by the **selectivity coefficient** $K_H^{M/n}$ which is also referred to as the **concentration exchange constant:**

$$K_H^{M/n} = \frac{[M^{n+}]_{/R/}[H^+]}{[H^+]_{/R/}[M^{n+}]} \tag{5}$$

where /R/ is the resin phase. The selectivity coefficient expresses the selectivity for the ions of M^{n+} in relation to the hydrogen ions from the solution containing the equivalent concentration of both these ions. As a suitable example, one can mention the selectivity of certain ions from a Dowex ion exchanger:

$K_H^{M/n}$ For univalent ions: $Li<H<Na<NH_4<K<Rb<Cs<Ag$

For bivalent ions: $Mg<Zn<Co<Cu<Cd<Ni<Mn<Ca<Sr<$
$<Pb<Ba$

For trivalent ions: $Cr<Ce = La$

According to this sequence, the sorption capacity of the ions by the given ion exchanger decreases.

The distribution coefficient of ion exchange K_I is the ratio of the number of ions in the resin $C_{M/R/}$ in grams to the number of the ions in the solution $C_{M/N/}$ expressed in ml:

$$K_I = \frac{C_{M/R/}}{C_{M/N/}} \tag{6}$$

In exchange of ions with different charges, the distribution coefficient depends greatly on the concentration of the external solution and increases with dilution of the solution. For example, in the exchange of a bivalent ion by an univalent ion, the concentration of the univalent ions increases and the univalent ion enters the resin and displaces the bivalent ion. In the case of a cation exchange resin, this favourable situation makes it possible to regenerate the ion exchange resin by hydrogen or sodium ions after saturation of the resin with calcium or magnesium ion. This takes place in, for example, softening or deionisation of water.

If a certain amount of an ion exchange resin populated by the ions A is introduced into the solution of an electrolyte containing ions B, a certain proportion of the ions A in the resin is substituted by the ions B and the same amount of the ions B in the solution is substituted by the ions A:

$$A_{/R/} + B_{/N/} \rightleftarrows A_{/N/} + B_{/R/} \tag{7}$$

/R/ is the ion exchange resin, /N/ is the solution.

The graphical representation of ion exchange is the exchange isotherm, Fig. 3, which is the dependence of the equivalent fraction of the ions A in the ion exchange resin on the equivalent fraction of the ions A in the solution. The equilibrium exchange isotherm specifies the proportion of one of the two competing ions as a function of the proportion of the equivalent of this ion in the solution. In this case, the division factor for the selective condition is given by the ratio of the areas I and II that come into contact at the appropriate point of the isotherm.

The efficiency of ion exchange, i.e. the degree of purification of the solution and the rate of the process, depends on the properties of the ion exchange resin, the condition of saturation with the counter-ions, the grain size, the composition of the supplied solution, the size of the column, the flow rate of the solution, and temperature. The best results are obtained with a low flow rate of the solution from the column, the smaller grain size of the ion exchange resin, the low concentration of admixtures in the solution, high temperature of the solution, a high value of the ratio of the height of the column to its cross-section.

Ion exchange is used in the following applications in the metallurgy of pure substances and also environmental protection:

- the deep purification of the solution of salts of refined elements to remove admixtures;
- the separation of two elements, with one of the elements remaining in the solution and the other element adsorbed by the ion exchange resin (Zr–Hf, Ta–Nb, Pr–Nd, separation of lanthanides, actinides,);
- concentration or obtaining an important element in ion exchange resin from a large volume of the diluted solution (Au, U);
- preparation of deionised and demineralised water for semiconductor technology;
- removal of harmful substances from industrial effluents;

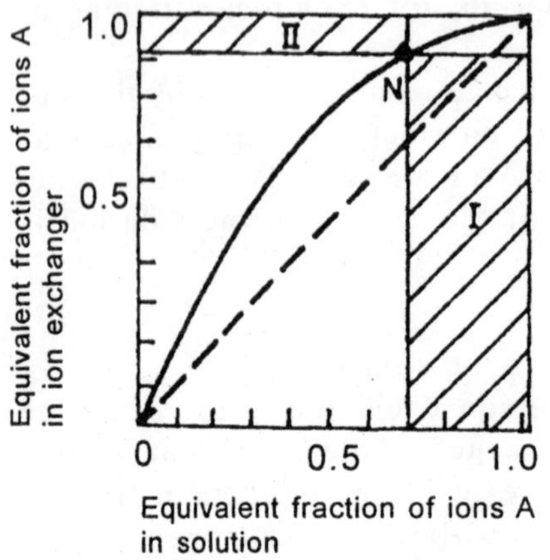

Fig. 3 Ion exchange isotherm.

- extraction of the required ions of metals from waste industrial solutions.

3.4. Chromatography

Chromatography also belongs to sorption processes and is used for the deep purification of substances in metallurgy, chemical technology, analytical chemistry, biology, etc. All the chromatographic methods are based on the different speed of movement of the components of the mixture through a porous sorption medium. The components to be separated are distributed between two phases, where one of the phases is stationary, is characterised by a large volume of surface, and forms the 'bed' of the chromatographic column, whereas the other phase is fluid, mobile, and passes through or along the stationary phase. The transfer of mass between the moving and stationary phases is caused by the fact that the molecules from the mixture are adsorbed on the surface of the particles or in pores, or transfer into the layer of the liquid on the surface or inside the pores.

The separation of the components of the specimens is based on the fact that the speed of movement of individual dissolved molecules through the column or a thin layer of the sorbent depends directly on the distribution coefficient of the components, which is the ratio of the equilibrium concentration of the components in the stationary phase to the concentration of the components in the moving phase. The distribution coefficient of chromatography of each component determines its content in the moving phase at any time and, consequently, the time during which the component remains in the stationary phase. This time subsequently determines the delay or a decrease in the speed of movement of the dissolved substance. If the differences in selective deceleration are significant, each component moves through the column or along the thin layer of the stationary phase at the speed which depends on its special sorption properties. If the components are separated, they leave the column at different times. The columns are filled with a fine-grained powder material of the same grain size, with the characteristic surface layer of the stationary phase which may have the following nature:

- adsorption – adsorption chromatography
- ion exchangers – ion exchange chromatography
- gel with the defined micropores – gel chromatography
- the thin liquid film – separation chromatography

The complete separation of the components of the mixture is obtained as a result of the so-called elution development of a chromatogram. A small amount of a mixture is placed in the upper part and a mobile phase is introduced into the column at a continuous, constant rate. This phase is represented either by a gas (gas chromatography) or a liquid – liquid chromatography. The mobile phase carries the individual components of the mixture at different rates, the components are delayed by the stationary phase with different efficiency, i.e. the values of the distribution coefficient of chromatography differ. From the column, the components leave more or less separated in the eluate (solution). The efficiency of separation of two substances increases with an increase of the difference of the elution time and volume and with a decrease of the weight of the elution peaks of the chromatogram, Fig. 4.

Adsorption chromatography is based on the application of the adsorbent as a stationary phase. In adsorption, intermolecular forces, for example, van der Waals forces, electrostatic forces between the dipoles, etc, operate.

The separation of polar substances is carried out using polar adsorbents, for example, silica gel or aluminium oxide, whereas the separation of non-polar substances is carried out using non-polar adsorbents, for example, active coal. Other adsorbents include cellulose, kaolin, talc, B_2O_3, graphite.

A suitable example is the process of chromatographic separation of mixtures of two components A, B of an aqueous solution, for example $CuSO_4$ with $CoSO_4$. The solution is passed through a glass column filled with an adsorbent, Al_2O_3, Fig. 4a. Since the copper ions are adsorbed by aluminium oxide more efficiently than cobalt ions, the copper ions will initially be retained in the adsorbent, whereas the cobalt ions will continue moving through the column, resulting in their separation from the copper ions. This is accompanied by the

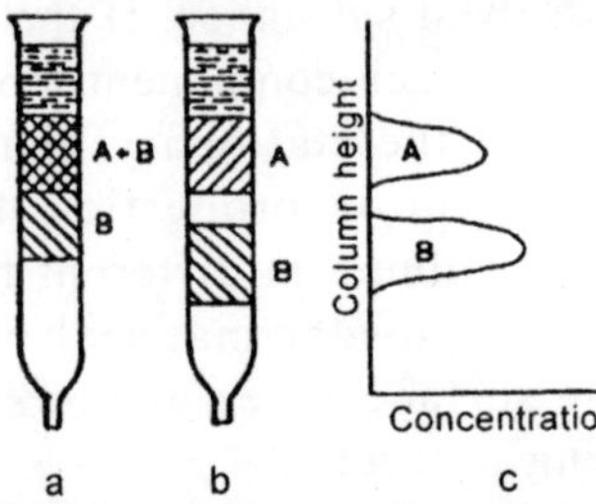

Fig. 4 Principle of chromatography: a) initial condition; b) distribution of components after chromatography; c) final chromatogram.

formation of two interfaces of movement of the ions: the leading interface of the cobalt ions and the lagging interface of the copper ions. This nonuniform distribution of the components along the height of the layer of the adsorbent results in the chromatogram shown in Fig. 4c. The adsorption affinity of these two molecules determines which of the two substances will be bonded more strongly on the surface of the adsorbent.

In desorption, the molecules of the dissolved substance are forced to move forward by the effect on the flow of the moving phase. This force acts on all components for the same period of time during which they are in the moving phase. Only the molecules with a high affinity for the adsorbent are selectively retarded. For complete separation, i.e., separation of copper from cobalt in the present case, a pure solvent (distilled water) can be passed through the column. In this case, the cobalt ions which are adsorbed far less efficiently washed-out from the upper part of the column and transferred into its lower part. As a result of passing a sufficient amount of the solvent, it is possible to separate completely and gradually wash out both types of ions.

In the presence of three or more components in the separated mixture, the chromatogram becomes more complicated. The primary chromatogram with three components consists of four bands, with the upper band containing all three components, the central band two components and the lower band only one component with the lowest adsorption capacity. In adsorption on aluminium oxide, cations are divided according to the following sequence: As^{3+}, Sb^{3+}, Bi^{3+}, (Cr^{3+}, Fe^{3+}, Hg^{2+}), Pb^{2+}, Cu^{2+}, Ag^{+}, Zn^{2+}, (Co^{2+}, Ni^{2+}, Cd^{2+}, Fe^{2+}), Ti^{2+}, Mn^{2+}, and the cations given in brackets cannot be separated. This series is independent of the anions present in the solution, e.g. SO_4^{2-}, Cl^-, NO_3^-, PO_4^{3-}.

In separation liquid–liquid chromatography, the components of the specimens are separated between the stationary liquid phase, trapped usually in the inert structure of the solid particles, for example, silica gel, quartzite, Al_2O_3, filter paper, powder cellulose, etc, and between the moving liquid phase. Both liquids are selected in such a manner that they are characterised by limited miscibility.

Because of the relatively high rate of separation, separation chromatography is used in a number of cases for the efficient separation of compounds, in particular organic compounds, phenols, amino acids, fatty acids, for the separation of unstable biological materials and compounds with a large relative molecular mass. The columns with sorbed tributylphosphate on silicone quartzite have been used, for example, for the separation of traces of arsenic from germanium, traces of scandium from calcium and also traces of nickel from molybdenum. Columns with tri-n-octylamine are used for the separation of traces of cobalt from nickel and traces of Fe^{3+} from manganese. In ion exchange chromatography, ion exchangers are used as the stationary phase. Both the pH of the solution and the structure of ion exchange resins, i.e. their selectivity, determine the speed of movement of cations or anions in the column. Therefore, with a change of the pH of the solution and the application of different liquid phases, such as chlorides, cyanides, organic solutions, it is possible to obtain individual cations or anions trapped at different heights of the column.

The methods of ion exchange chromatography are especially effective for the separation of elements with very similar properties, for example, lanthanides. A solution of the mixture of lanthanide compounds is passed through a column with ion exchange resins with a solution of citric acid. The individual cations of the lanthanides mutually differ only slightly by the degree of adsorption and, consequently, the speed of movement in the column. Therefore, in the ion exchangers they are gradually sorbed with the formation of individual zones along the height of the column. During elution, i.e. washing with the solvent, the separated solutions of eluated elements gradually leave the bottom part of the column with relatively high purity. Initially, the solution of the heaviest lanthanide with the lowest adsorption capacity is discharged, this is followed by solutions of the elements with lower atomic masses. Gas chromatography is used not only in the separation of easily volatile compounds such as, for example, inorganic halides of metals, chlorides of titanium, silicon, tantalum, niobium, zirconium, hafnium, etc, but also as a method in analytical chemistry. Sorption processes, i.e. adsorption, ion exchange, chromatography, are of considerable importance for the separation and deep refining of intermediate compounds especially in the primary hydrometallurgical stage, and these compounds are used in the second refining stage as initial materials for obtaining high-purity substances.

4. EXTRACTION

4.1. Physico-chemical nature of extraction and its importance

In a wider sense of the word, extraction is the process of selective leaching of compounds into the liquid phase. For any type of extraction process, it is important that the solvent is in the liquid condition and the mixture to be extracted maybe gaseous, liquid or solid. The extraction of the gas by the liquid is absorption. Extraction of solid compounds by liquid compounds is leaching used widely mainly in hydrometallurgy. The extraction of a dissolved compound from one liquid phase to another is liquid extraction or also liquid–liquid extraction, or extraction from liquid into liquid. This method is of considerable importance in the technology of separation and production of pure substances [3].

Liquid extraction is a process of leaching the dissolved substance from one liquid solvent by another liquid solvent which is not miscible with it. The liquid from which the dissolved substance is extracted is in most cases an aqueous solution. The liquid used for the extraction of the dissolved substance is the extrahent. The extraction agent is usually an organic phase which is formed either by a pure solvent or contains a reagent represented by a component with chemical reaction with the leached substance, and a diluent which is part of the extraction agent, for example, for decreasing viscosity. In a number of cases, the solvent may be a reagent and the diluent in the process of extraction is not always inert in relation to the substance being extracted. The component which is soluble in both phases is separated between them at a specific ratio. Equilibrium is obtained when the free energy of the component in both phases is identical.

Liquid extraction is used for:

- separation of a mixture of two substances with concentration of each mixture in a different solvent (aqueous and organic);
- extraction of the main component of the mixture into the organic solvent with the admixture left in the aqueous solution, or vice versa;
- extraction of the admixture into the organic solvent with the main component remaining in the aqueous solution.

Extraction is a rapid, relatively simple and convenient process and is used widely. Extraction separation may be used for both very low (trace) concentrations and high concentrations of elements in the solution. The individual components can be selectively extracted from aqueous solutions into organic solvents or extracted from the organic into aqueous phase. The required efficiency of separation is obtained by selecting suitable experimental conditions which are: pH of the solutions, the type of masking agent ($CaCl_2$), the solvent and the extraction agent.

4.2. Mutual solubility of two liquids

There are liquid systems whose components are mutually insoluble (non-miscible), such as, for example, water–oil, water–mercury, and others, and but there are also liquid systems those components are characterised by limited solubility, for example, water–aniline, water-phenol, etc. However, many systems are characterised by complete mutual solubility of the components, for example, water–alcohol.

The mutual limited solubility of two liquids is observed in systems in which the formation of a solution is accompanied by the absorption of a large amount of heat. The change of the external conditions, in particular, temperature, affects the mutual solubility of the liquids. An increase of temperature increases solubility, if dissolution is accompanied by the absorption of heat, otherwise, it decreases solubility.

Figure 5 shows the temperature–composition diagram for a phenol–water system characterised by mutual limited solubility of the components. In the diagram, the points P correspond to the saturated solution of phenol in water, and the points Q to the saturated solution of water in phenol. Since the concentration of the saturated solution of one component in another increases with increasing temperature, the points P and Q will become closer during heating until finally at the so-called critical dissolution or mixing temperature (point K) they become identical. The region situated inside the QKP curve is heterogeneous and consists of two phases. The points outside the region represent a homogeneous system; in this region the components are characterised by complete miscibility. The lines connecting the points P_1P_2K and Q_1Q_2K is referred to as the solubility curve, the liquation curve or also the binodal curve. At low temperatures, the system is heterogeneous and divided into two phases: phase Q with a high phenol content (the phenol phase), and phase P with a high water content (aqueous phase).

Heating of the system is accompanied by the change of the

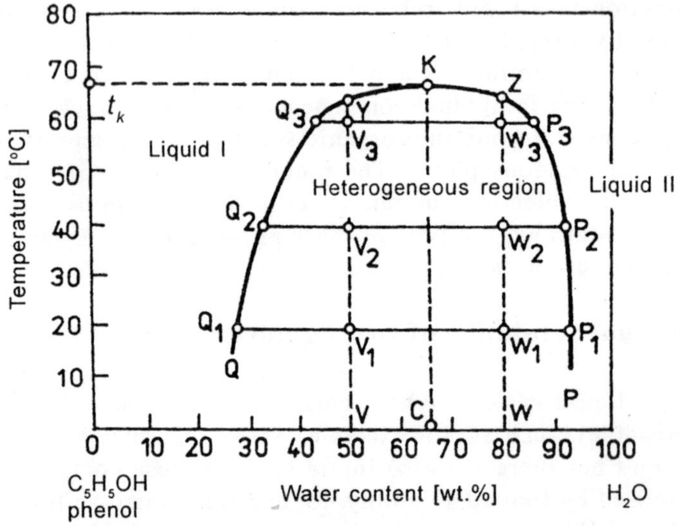

Fig. 5 Temperature–composition diagram of the phenol–water system.

composition of the phenol phase along the curve $QQ_1Q_2Q_3$ and the composition of the aqueous phase changes along the curve $PP_1P_2P_3$. In this case, the phase Q is in equilibrium with phase P at the given temperature. In the phenol–water system for composition V, the temperature corresponding to the point Y on the binodal curve is the minimum temperature of complete mixing. For composition W, this point is represented by Z.

During gradual heating of the solution with composition V the amount of the aqueous phase will decrease and at point Y the phase disappears and the system becomes homogeneous. The same can be mentioned for gradual heating of the solution with composition W; the amount of the phenol phase will decrease and at point Z the phase disappears and the system becomes homogeneous. The direction of the process is controlled by the position of the point corresponding to the composition of the system, i.e. to the right or left of the CK connecting line.

The problem of the mutual solubility of the liquids or of their insolubility is of considerable importance for liquid extraction. The compounds of similar composition, structure and size of the molecules are usually characterised by good mutual solubility. In extraction, the molecules of the extracting solvent do not have to be polar, if the polarisation bond is typical for the molecules of the solvent from which the extract is obtained, and vice versa. For example, if extraction takes place from an aqueous solution, the other solvent (extrahent) must be a non-polar liquid of organic origin. The main requirement for the extrahent is chemical inertness in relation to water and very low solubility in water. The extracted substance must be dissolved in both liquids, more in one and less in the other. This is always associated with well-known difficulties because if the given substance is characterised by high solubility in one of the contacting phases (for example, polar), it is then characterised by very low solubility in the other phase (non-polar).

If a substance is characterised by high solubility in a polar solvent, the molecules of the substance must also be polar. However, this substance will be characterised by low solubility in a non-polar liquid. For example, the acetates of copper and calcium and also salts of inorganic acids of these metals are characterised by high solubility in water. However, they are not sufficiently extracted by means of hydrocarbons because the salts are polar.

These difficulties may be eliminated if the dissolved metal is transferred into a compound which has both polar and non-polar group. Complicated compounds assume solubility of the main part (polar or non-polar). For example, the salts of copper and other salts of high-molecular fatty acids are capable of extraction from the aqueous medium by hydrocarbons, and the extraction capacity increases with increasing length of the hydrocarbons chain, i.e. with increasing non-polar part. Therefore, for example, copper acetate is not extracted and the copper salt of oleic acid is extracted incompletely. Also, the salt of valeric acid is extracted far more efficiently, whereas the copper salts of higher fatty acids are extracted almost completely.

4.3. Separation laws

If a small amount of the third component is introduced into the system of two non-mixing layers, then after establishment of equilibrium this component appears in both phases in a specific concentration. If the concentration of the third component is low, the size of their fractions in both phases is the same. An increase of the amount of this component in the system results in a proportional increase of its concentration in both liquid phases.

The separation constant K_D for a pair of ideal solutions in extraction is given by the ratio of molar fractions of the dissolved substance in the organic $x_{/p/}$ and aqueous phases $x_{/N/}$:

$$K_D = \frac{X_{/p/}}{X_{/N/}} \tag{8}$$

The separation constant is independent of the total actual concentration of the component in the individual phases, provided one or both phases are not saturated with this component. If more than one component is separated through the phase boundary (at low concentrations of these components), then according to the Nernst law of independent separation it applies that each component will be separated independently of other compounds.

For every given temperature, the ratio of the concentrations of the third component in two equilibrium liquid phases is constant. This is the principle of the separation laws. For example, let us consider the water–carbon disulphide system where H_2O and CS_2 are mutually insoluble. If an iodine crystal is introduced into this two-component system, then when equilibrium is established the iodine will be present in both layers. However, its concentration in carbon disulphide will be approximately 600 times higher in comparison with that in water. However, if different amounts of iodine are introduced into the system, the ratio of the concentrations of iodine in one of the layers to the concentrations in the other layer will remain constant at constant temperature, and when introducing small amounts of iodine:

$$K_{DI} = \frac{X_{I/CS_2/}}{X_{I/H_2O/}} \tag{9}$$

where $X_{I/CS_2/}$ is the concentration of iodine in carbon disulphide.

However, in the concentrated solutions of the third component, the value of K_D does not remain constant. The introduction of certain quantities of the third component changes the mutual solubility of the first two phases and this complicates the relationship between the composition of the equilibrium liquid layers and may contribute to the formation of a homogeneous or a three-layer system.

The separation ratio D_C of a substance between the organic and the aqueous phase is given by the following equation

$$D_C = \frac{C_{A/org/}}{C_{A/H_2O/}} \tag{10}$$

The numerator contains the total concentration of component A in all its chemical forms in the organic phase and the denominator is the total concentration of component A in all chemical forms in the aqueous phase. The efficiency of extraction depends on the separation ratio D_C and also on the volume of the individual phases and the number of extraction processes only if the separation ratio is high.

If the volume of the aqueous phase is V, the volume of the organic phase V_{org} and the original concentration of the component in the aqueous phase is C_0, the concentration is given by the following equation:

$$C = C_o \frac{V}{D_C V_{org} + V} = C_o \frac{1}{1 + D_C \frac{V_{org}}{V}} \tag{11}$$

The concentration of the component remaining after single extraction (shaking) depends evidently on two factors: the separation ratio D_C and the ratio of the volumes of the two phases. A component can be removed by extraction to the required extent by repeated shaking or by the method of continuous extraction. After n extraction cycles, the concentration of the component remaining in the aqueous phase is given by the following equation:

$$C_n = C_o \left[\frac{V}{D_C V_{org} + V} \right]^n \tag{12}$$

The separation into the individual phases is the result of attraction and repulsion exerted by the molecules or ions of the competing liquid phases on the dissolved substance and on each other. The following physical interactions are important:

- dispersion interaction
- dipole–dipole interaction
- induction interaction
- hydrogen bond

4.4. Isothermal equilibrium in a ternary liquid system

The concentrated solutions containing three components can be examined more efficiently using a ternary diagram indicating the composition of each component of the system (Fig. 6). In order to obtain information on the distribution in ternary systems, attention will be paid to investigating the process of addition of substance A into a heterogeneous mixture of components B, C.

It is assumed that the components A and B and also A and C mix together completely, and the components B and C are characterised by limited mutual solubility. All binary systems of the components B and C, whose composition is indicated by the points distributed between P and Q (Fig. 6) are divided into two phases with the

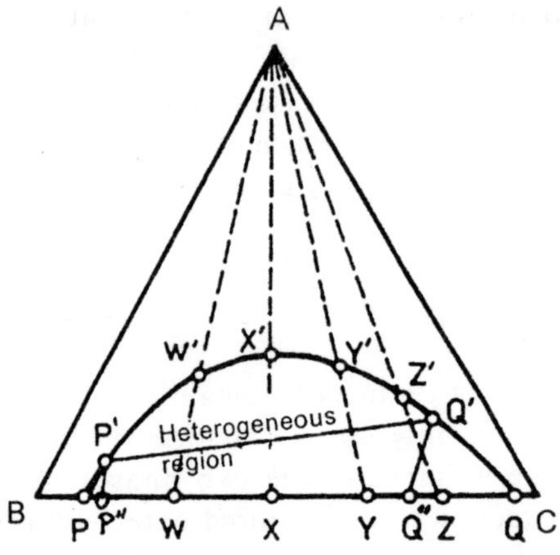

Fig. 6 Binodal curve in the ternary diagram of the A–B–C system.

composition corresponding to P and Q. We examine the system whose general composition is given by W. This mixture is divided into phases P and Q with ratio WQ:WP in accordance with the lever rule. After adding component A to the system W, the ratio the phases WQ:WP gradually changes. The amount of one of the phases decreases in until it reaches the composition at which the system W contains the amount of component A which is such that the phase with the high concentration completely disappears. The system will be homogeneous.

This condition is indicated in the diagram by W' which is the mixing or saturation point in the A–B–C ternary diagram. The binodal curve PW'X'Z'Y'Q in the ternary diagram for the given temperature is the solubility isotherm. In the region restricted by the straight line PQ and the binodal curve PW'X'Z'Q, the system is always heterogeneous (two-phase region), and in the remaining part of the diagram the system is homogeneous. In the heterogeneous region, any system will be divided into two coexisting phases whose composition is indicated by the points distributed on the binodal curve.

In the temperature–composition diagram for the binary extraction system, the coexisting phases are situated on the horizontal line, i.e. parallel to the axis x, because both binary phases are characterised by the same temperature, for example, the connecting line Q_2P_2 in Fig 5. However, in the ternary diagram, this line is usually not horizontal. The angle of the connecting line changes depending on the nature of the component. As an example, Fig. 6 shows the line segment P'Q' as the line connecting the coexisting phases.

As mentioned previously, in the case of a low concentration of the third component A, its distribution constant K_{DA} in the system of two liquid non-miscible phases has a constant value. The distribution constant represents the ratio of the concentration of this component Q' and P' and this can be exposed by the following equation:

$$K_{DA} = \frac{\text{concentration A in phase Q'}}{\text{concentration A in phase P'}} = \frac{QQ''}{PP''} \quad (13)$$

However, in some cases, the distribution constant may also have the values K_{DA} = 1, or K_{DA}<> 1. Therefore, for the case shown in Fig.6, K_{DA} > 1.

4.5. Principle of extraction from solutions in the ternary system

A mixture of two liquids A and B with almost complete miscibility can be divided in many cases into components by means of extraction. For this purpose, a solvent (extrahent) is added to the mixture. The solvent does mix completely with a mixture and is capable of dissolving one of the components to the required extent. In the ternary diagrams, the pure solvent S is usually placed in the right bottom corner of the triangle, component B, characterised by limited solubility in the solid, is placed in its left corner, and component A, characterised by complete solubility in the solvent is placed in the upper tip of the diagram (Fig. 7).

With gradual addition of the solvent S to the mixture of A and B (indicated by point F in Fig. 7), the point, corresponding to the total composition of the ternary system, moves on the straight line FS. If one considers the same part of the solvent and the mixture F, the total composition corresponds to the point M_1, and $M_1S = FM_1$. Point M_1 represents a heterogeneous mixture which is divided into two phases after the establishment of an equilibrium. The composition of these two phases indicated by the points P_1 and Q_1. Phase Q_1 is characterised by a high content of the solvent and is usually referred to as the extracted phase, the extract, or also the phase of the solvent. Phase P_1 with a low content of the solvent is the raffinate of the residual phase.

The precipitation of the solvent from the extract Q_1, which takes place in either by distillation or rinsing, results in the formation of an extract without the solvent, whose composition is indicated by the point E_1 in Fig. 7. The same procedure is used to obtain the raffinate R_1 (residue) from the raffinate phase P_1. Using this method, extraction, with subsequent dissociation of the phases and the removal of the solvent from them, is used for the separation of the initial mixture F into the mixtures E_1 and R_1. In comparison with the initial mixture F, the mixture E_1 is characterised by the higher content of the component A and the mixture R_1 by the lower content of the component A. In the case of the component B the ratio is reversed. Consequently, single stage extraction is characterised only by the partial separation of the components from the mixtures A and B.

Further addition of the extrahent S to the raffinate R_1 or to the phase P_1 results (after separation of the phases and the removal of the extrahent) in the formation of the new raffinate R_2 with an even higher content of the component B in comparison with R_1. Using this method, it is possible to obtain, from the initial mixture, the component B of almost any degree of purity after multistage extraction.

As regards the extract, containing component A, the maximum attainable degree of purity of this component is indicated by the point E_m (Fig. 7), where E_mS is the tangent from the point S to the binodal curve.

In order to increase the purity of the extract, i.e., to produce it with the highest possible content of component A, the extracted product is subjected to additional treatment with the so-called rinsing solution. This solution, which does not dissolve the component A, must have the capacity of dissolving efficiently the component of the raffinate B which contaminated the extract. This results in the formation of a high-purity extract.

The results of extraction depend on the following factors:

– the size of the two-phase region;

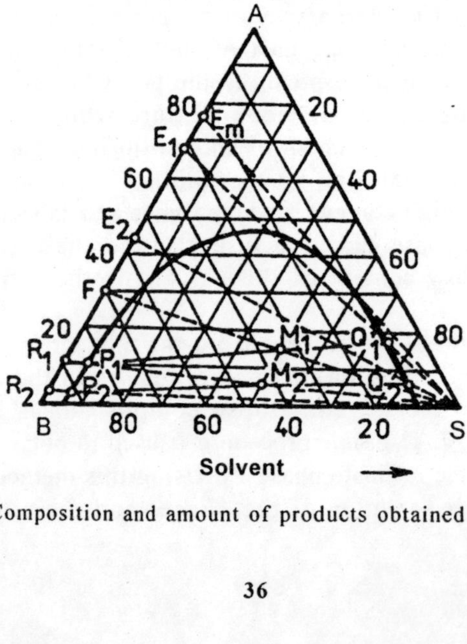

Fig. 7 Composition and amount of products obtained in extraction.

– the angle of the line connecting the coexisting phases, i.e. on the value of the distribution ratio;
– on temperature;
– on the nature of the solvent–extrahent.

4.6. Extraction methods

The dissolved substances are usually transferred from one liquid solution to another one because it is necessary to separate or purify the required component. In some cases, the required component is removed from the initial solution in which the undesirable compounds are then retained. In other cases, the required solution remains in the original solution and impurities are extracted. Extraction can also be used for producing another solution from which the required component can be removed more easily in comparison with the original solution. Shaking is used when it is possible to obtain high values of the distribution ratio of the required component. Subsequently, the efficient removal of the required component requires only a small number of extraction cycles by shaking. For example, if the distribution ratio is higher than 4, three extraction cycles are sufficient in order to remove the component to 99%. Shaking is usually carried out using a separation funnel which is pear-shaped. Both phases are efficiently shaken in the funnel, the phase is subjected to gravitational separation by settling and, subsequently, the phase is discharged through a stopcock. In the case of a low distribution ratio and a high separation factor:

$$a_D = \frac{D_{c_1}}{D_{c_2}} \qquad (14)$$

where D_{c1} and D_{c2} are the distribution ratios of both components, and in this case the continuous extraction methods are efficient. These methods are based on the continuous flow of the non-miscible solvent through the extracted solution. If the solvent is volatile, it can be eliminated and returned back by distillation and condensation. Despite the fact that in the short period of time in which the two phases are together in contact it may not be possible to obtain the distribution equilibrium, the dissolved component is continuously removed by the extracted phase. The statistical parameters of the process of liquid extraction are usually characterised by the number of extraction steps required for obtaining the necessary efficiency of separation.

The extraction stage is every stage of processing the liquid containing the extracted compound, by the new amount of the extracting agent. The stage is equilibrium if extraction results in the equilibrium distribution of these compounds between the contacting liquids. On the whole, the process of liquid extraction includes the following stages:

- the transfer of the extracted compound in the compacted (aqueous) phase to the surface of the droplets of the dispersed phase (extracted compound);
- reaction between the extracted compound and the dissolved substance on the surface of the phase boundary;
- the transfer of the extrahent inside the droplets of the dispersed phase.

The limiting stage of the process is the diffusion of the dissolved substance in the aqueous phase. Consequently, the process of liquid extraction is efficient in the case of intensive mixing which greatly accelerates the diffusion transfer of the component. The rate of extraction depend strongly on the size of the disperse phase: the small size of the droplets results in more rapid establishment of extraction equilibrium. The efficiency of separation depends on the value of the distribution ratio, the viscosity of the phases, the ratio of the volume of the phases, the area of their mutual contact and their mutual velocity. During the passage of the extrahent through the solution, it is possible to use stops or mixing devices in order to obtain closer, more efficient and longer contact of the two phases. Equipment for continuous extraction is based on the distillation of the extracting solvent from the boiling flask, condensation of the solvent and the passage of the condensate through the extracted solution.

The process of liquid extraction may be periodic, continuous or of the counter flow nature. Periodic extraction is shown schematically in Fig. 8a. The extracting agent S is added to the mixture F to be separated and is mixed together with this mixture. After establishment of equilibrium between the compacted and dispersed phases and separation of the phases, the phase of the raffinate P_1 is again processed by the extrahent S, and so on. These operations are repeated several times. Finally, from the phase of the raffinate phase (for example, P_3) leaving the last device (extractor) the extrahent is obtained by distillation resulting in the formation of a product with a high content of component B in or a low content of component A.

In continuous liquid extraction, the components are mixed with the settling of the phases taking part in the different extracting devices or extracting columns, Fig 8b. The transfer of mass between the initial solution and the extrahent takes place in the mixers M_1, M_2 and M_3, and the separation of the phases in the settling containers Z_1, Z_2 and Z_3. The initial solution F and extrahent S are continuously supplied

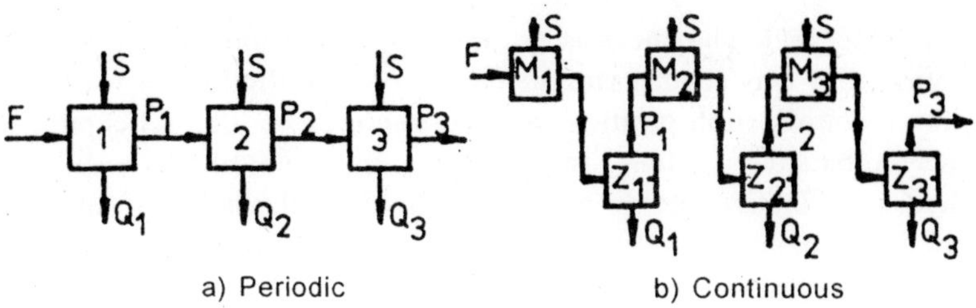

Fig. 8 Scheme of liquid extraction.

into the mixer M_1 in which they are mixed together. The resulting heterogeneous mixture is continuously added to the settling containers Z_1 in which the phases are separated. The raffinate phase travels from Z_1 to the next mixer M_2 into which the solvent is also added (continuously). The heterogeneous mixture which forms in the mixer M_2, is continuously transferred into the settling container Z_2 which operates on the same principle as Z_1, etc. In each settling container, the extract phase Q_1, Q_2, Q_3 is produced. The final phase of the raffinate P_3 from the last settling container is transferred to the distillation column in order to remove the extracted substance and obtained the pure refined product.

The application of the separation methods developed for the liquid-liquid system can also be extended to the separation of substances whose separation constants differ only slightly. This is possible using the so-called counter flow extraction which is a multiple separation process with a large number of steps, perfectly discontinuous, and consisting of stages. It requires equipment with 100, 200 or even 400 separation chambers in which extraction and transfer of the phases can be carried out automatically. Equipment operates in such a manner that both phases are automatically mixed and one of the phases (moving) is transferred into the next chamber. The substances with a high value of K_D are characterised by the highest speed and are followed by the compounds with lower values of K_D. This method is used mainly for the separation of complicated mixtures of organic compounds.

4.7. Extraction systems

The simplest methods of extraction of metallic ions are based on the formation of neutral, hydrophobic molecules. These reactions may take place either in the aqueous phase or at the phase boundary. Important agents used in this case include chelates, mono- or diesters of phosphoric acid, carbonic or sulphonic acid distributed in an inert organic solvent. The chelates are important compounds for extraction because they may compensate the charge of metallic ions or may occupy its own coordination position (neutral chelates or internal complexes). These substances include, for example, enols of diketone, dithizone, 8-hydroxiquinoline, etc. In particular, the d-2-ethylhexyl of phosphoric acid is an excellent extraction agent for a number of metals such as lanthanides and actinides.

The next group of extraction systems enclose the formation of solvents in organic dissolvents. This is accompanied by the formation of donor and acceptor compounds whose strength depends on the basicity of the organic solvent. Suitable examples of these types of solvents include ether, alcohols, esters of carbonic acid, ketones and aldehydes. This method may be used for the extraction of complex acid of metals by ether, for example $H[FeCl_4]$. Recently, the extraction agents containing phosphorus, for example, trialkylphosphinoxide R_3PO, alkyldialkylphosphonate $(RO)R_2PO$, dialkylalkylphosphonate $(RO)R_2PO$, trialkylphosphate $(RO)_3PO$ have been used. One of the most important extraction agents of this type is tributylphosphate $(C_4H_9O)_3PO$, denoted as TBF. It is characterised by high volatility, high resistance to mineral acid and efficient mixing with organic solvents. The following reactions take place during extraction from nitrate solutions:

$$M^{3+} + 3\ NO_3 + 3\ (TBF) = [M(NO_3)_3.3\ TBF] \quad (15)$$
$$MO_2^{2+} + 2\ NO_3 + 2\ (TBF) = [MO_2(NO_3)_2.2\ TBF] \quad (16)$$

Tables 3 and 4 give several examples of extraction separation.

4.8. Application of extraction in the metallurgy of pure substances

Extraction is the first step of separation and refining of compounds because the resultant purity is not very high. It is used in the hydrometallurgical production of metals where a metal is transferred from solutions containing also a large number of compounds of several metals selectively using suitable extraction organic agents into the organic phase, i.e. the phase which does not mix with the remaining aqueous phase. After separation of both phases, the metal is transferred from the organic phase using, for example, an acid, back to the aqueous phase and this is accompanied by the regeneration of the applied organic solvent. Recently, this technology has been used for the

Table 3 Some extraction systems for analytical applications

Extracted ion	Complex compound Extraction agent	Solvent
Fe^{3+}	Thiocyanatane (CNS^-)	Ether
Al^{3+}, Mg^{2+}, Fe^{3+}	8–hydroxiquinoline	Chlorophorm
Th^{4+}, Cu^{2+}	Diacethyldioxim	Chlorophorm
Ni^{2+}, Pd^{2+}, UO_2^{2+}	4–(2–pyridylazo)–resorcin	Ethylester of acid
Mn^{2+}, Cu^{2+}, Pb^{2+}	Diphenylthiocarbazon	Carbon tetrachloride
Fe^{3+}, Co^{2+}, Ni^{2+}	Ammonium tetramethylene dithiocarbamate	Chlorophorm
Rare earth Actinides	di-2-ethylhexyl of phosphoric acid	n-heptane
Zr^{4+}, Ti^{4+}	Tributylphosphate	Tributhylphosphate
Fe^{3+}, Co^{2+}, Cu^{2+}	Trioctylamine	Benzen

separation of a number of metals, U, Th, Zr, Hf, Nb, Ta, Cu, Zn, Ni, Co, lanthanides, actinides, etc.

The chlorides are the simplest and cheapest salts and a number of available solvents can be used for their extraction. Consequently, systematic investigations have been carried out into the distribution ratios of chlorides of a large number of metals mostly characterised by high solubility in water or in oxidised and salt solutions. In the majority of cases, isoamylalcohol (i-$C_5H_{11}OH$) was used as a sufficiently universal solvent of the mineral salts, and $CaCl_2$ was used as the salt soluble in water.

The extraction from liquid into liquid is one of the relatively efficient processes of obtaining, separating and purifying compounds. It is usually carried out as the counter flow multistage separation process and it is highly promising in particular for the production of a number of selectively operating solvents, complex making agents, separators of salts, i.e. systems with high values of the separation factor.

Table 4 Examples of extraction separation

Element	Origin of sample	Composition of aqueous phase	Composition of organic phase
Ag	Concentrates of Pb, Cu, Au	8N–H_2SO_4	DZ in C_6H_6
As	Pyrite	H_2SO_4, KI, Na_2SO_3	Xantogenan in CCl_4
Au	Electrolytic sludge	6N–HCl, HNO_3	Ether
Be	Fission products	EDTA, pH 7–8	Acetylacetone in $CHCl_3$
Bi	Alloys of Pb, Sn	HNO_3, thiourea	$CHCl_3$
Cd	Zn	0.1 M–KI, H_2SO_4	Isopropylether
Ce	Fission products	pH 2.7–3.0	TBF + TTA in $CHCl_3$
Co	Ni	HCl	TBF
Cu	High purity materials: Al, Fe, Sb, Si, Sn, Zn	DDC, pH 7	$CHCl_3$
Ga	Bauxite	6N–HCl	Ether
Ge	Minerals, ash	9N–HCl	CCl_4
In	Ores	18N–H_2SO_4, 2N–HBr	C_6H_6
Mo	Minerals	$Na_2S_2O_4$, tartrate	Isopentyl-acetate
Ni	Alloys	Diacetyldioxim in ethyl alcohol	$CHCl_3$
Sb	Sn, Pb	Rodamin B	Isopropylether
Se	Ores	EDTA	Toluene
Sr	Fission products	pH 8	TTA–MIBK
Ta	Ores	HF	Methylorange in C_6H_6
Th	Monazite sand	PAN, pH 6.7–6.9	Ethyl acetate
Th	Ores	HNO_3, $Al(NO_3)_3$	Hexane–TBF
U	Lean ores	SCN^-, EDTA, HNO_3	TBF
U	Alloys of Zr	HF, $Al(NO_3)_3$	TBF
Zr	Steel	pH 3–8	TTA in xylene
Zr	Spent nuclear fuel	$HClO_4$, H_2O_2	Acetylacetone in $CHCl_3$

Abbreviations: DDC – diethyldithiocarbamate, DZ – dithizone, EDTA – chelaton 3, MIBK – methylisobutylketone, PAN 1–2–pyridylazo–2–naphtol, TBF – tributylphosphate, TTA 2-thenoyltrifluoracetone.

5. CRYSTALLISATION

The crystallisation refining methods are based on the different solubility of the admixtures in the liquid and solid phases. In the production of pure metals and semiconductor materials, crystallisation is used in two consecutive purification stages:

- crystallisation from aqueous solutions – selective crystallisation in the first stage of purification;
- crystallisation from melts – the most important method of the second stage of purification and final refining of metals and semiconductor materials [1].

In crystallisation from solutions, the metal salt is transferred from the solution into the crystalline phase. Crystallisation takes place from the supersaturated solution. The admixtures which have been transferred into the solution together with the refined substance are present usually in the solution at low concentrations away from saturation and, consequently, do not crystallise and remain in the solution. The situation in which it is necessary to carry out separation of two or more substances present in the solution in higher concentrations is more complicated. Even in this case, it is possible to carry out their separation on the basis of their different solubility by the method of selective crystallisation followed by recrystallisation of the already separated salts.

5.1. Crystallisation from solutions. Equilibrium, mechanism and kinetics of crystallisation

If a dissolved substance is brought into contact with water, it is gradually transferred into the solution. This process continues until the concentration of the substance in the solution reaches a specific value which is independent of the amount of the solid substance. The equilibrium is established between the solid substance and the solution. The solution becomes saturated. The disruption of this equilibrium with the precipitation of crystals from the solution takes place if temperature is reduced or if a part of the solvent is removed as a result of evaporation. The solution becomes supersaturated in relation to the dissolved substance and after establishment of equilibrium this substance crystallises with precipitation in the solid phase. This process continues until the solution becomes again saturated at the given temperature and equilibrium is established between the substance and its solution.

The crystallisation process consists of two stages:

- the formation of nuclei of crystallisation with a decrease of temperature as a result of the disruption of phase equilibrium;
- the growth of the resultant nuclei to large crystals.

The mechanism of the formation of nuclei may be described as follows: in the saturated solution, the molecules of the liquid are in continuous and disordered motion and possess a specific internal energy for the given temperature. The solution is characterised by the formation of clusters of molecules with a similar spatial arrangement as in the solid substance. As a result of thermal motion, these clusters of the molecules either form or breakup out at the same rate. With a decrease of energy, i.e. with a decrease of temperature, the clusters become more stable. They meet more frequently and, consequently, increase in size. This facilitates the formation, in the solution, of crystalline nuclei as a source of actual crystallisation.

The growth mechanism of crystals is based on adsorption, surface diffusion and the growth of steps on the surface of the nucleus. The growth of crystals is reflected as a regular process of growth of new areas. The rate of growth of the crystals is determined by the rate of diffusion transfer of the substance from the volume of the solution to the surface of the growing crystal and depends on the degree of supersaturation and on temperature.

5.2. Solubility of salts

For refining separation of substances (metal salts) by crystallisation from solutions it is necessary to know their solubility in water. The strongest effect on the solubility of solid substances is exerted by temperature. The increase of solubility with increasing temperature may take place in cases in which the dissolution process occurs simultaneously with the absorption of heat. The decrease of solubility with increasing temperature is observed when dissolution takes place with the release of heat. If dissolution is not accompanied by a thermal effect, the change of temperature will have no effect on solubility, i.e. solubility does not change with temperature. If the sign of dissolution enthalpy changes, the direction of the solubility curve will also change, i.e. it must have a maximum or a minimum of solubility.

5.3. Isothermal evaporation and crystallisation

For crystallisation from the solution to take place, there must be initially evaporation in order to produce a saturated solution from the non-saturated solution and subsequent cooling of the solution should result in supersaturation and crystallisation. The isothermal solubility curves are very important for the expected behaviour of the solutions subjected to isothermal evaporation.

If an unsaturated solution of two simple salts in the equimolar quantity, for example, point X in Fig. 9, evaporates at a constant temperature, the component A whose equilibrium straight line is intersected by the straight line OD, is the first to be separated. Consequently, the solution is enriched in component B. Further evaporation results in further precipitation of component A and the composition of the solution changes, as indicated by the arrow, until the composition of point C is reached. This indicates the start of precipitation of both component A and B and the composition of the solution remains constant. The resultant of evaporation is a mixture of both components.

5.4. Cyclic separation of salts by crystallisation from solutions

The separation of two salts in practice is carried out on the basis of different solubility of the salts with a change of temperature. As a result of gradual evaporation at high temperature and cooling to low temperatures it is often possible to obtain sufficient separation of salts. This cyclic process will be demonstrated on the example of the simple NaCl–KCl system, Fig. 10, which does not result in the formation of a binary salt and in which the individual salts crystallise separately at all temperatures higher than 0 °C. In Fig. 10, the curves *ac* and *cb* represent the solubility curve for NaCl and KCl at a

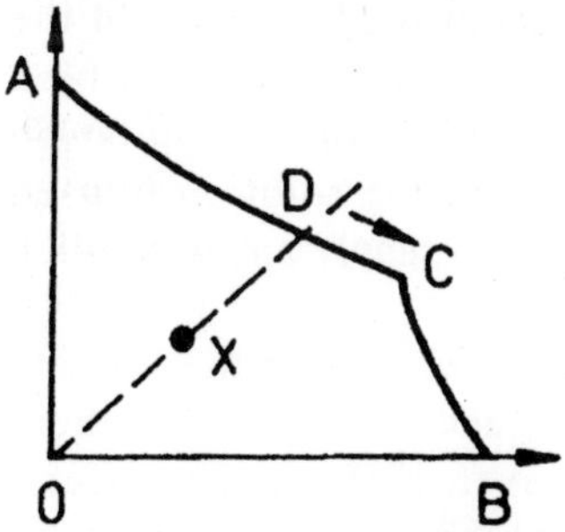

Fig. 9 Diagram of isothermal evaporation.

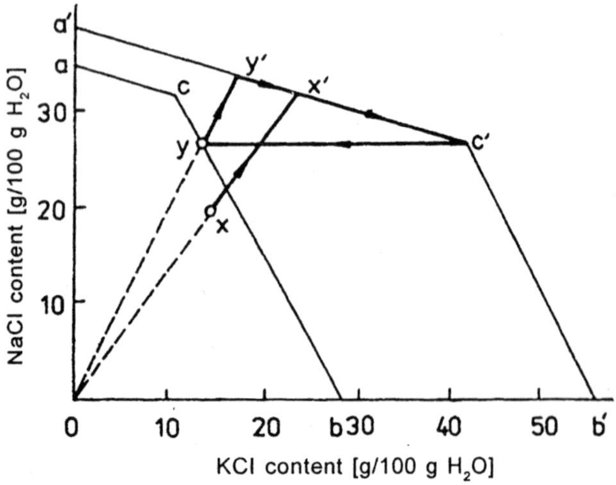

Fig. 10 Separation NaCl and KCl salts by means of the cyclic crystallisation process.

temperature of 0 °C, and the curves *a'c'* and *c'b'* are valid for a temperature of 100 °C.

As a result of evaporation of the initial saturated solution *x* at 100 °C, the composition changes along the curve *xx'*. With further evaporation, NaCl starts to crystallise from the solution and the composition of the solution changes in the direction *x'c'*. In the vicinity of point *c'*, the crystallised NaCl separates from the solution which is then cooled to 0 °C. During this process, the composition of the mixture changes in the direction *c'y*. Since the curve is situated in the region of potassium chloride and corresponds to its saturated solution, potassium chloride will precipitate until the solution reaches the composition corresponding to point *y* at a temperature of 0 °C. At this point solid potassium chloride is separated from the solution. If this is followed by cyclic evaporation of the solution, at composition *y'* NaCl starts to be separate and crystallisation continues until the point *c'* is again reached. Subsequently, the salt is separated and the isolated product is cooled down. This cycle of operations in which NaCl is the first to precipitate and is followed by the precipitation of potassium chloride can be repeated until all the solution is used up. In each evaporation cycle it is also possible to supply a new solution to the system.

In general, it may be said that if the ratio of two salts in the solution saturated by both salts at any temperature t_1 and t_2 is the same, there will be no separation of these salts. However, if this ratio differs, separation is possible only in cases in which the ratio increases sufficiently with temperature. For example, in separation of salts K_2SO_4 and KCl, the ratio $KCl:K_2SO_4$ at 0 °C is 23.9 and at 100 °C it is only 24.2. The difference is too small to result in an efficient process.

In separation of several salts forming a solution and characterised by different solubility in water, it is efficient to use fractional crystallisation based on the fact that different salts saturate the solution at different temperatures.

5.5. Freezing-out

Directional freezing-out is one of the specific processes of separation of aqueous solutions of salts. It is accompanied by the precipitation of crystals from a large volume of salt solutions (for example, seawater) whose melting point, depending on the concentration of salts in them, is several degrees centigrade lower than the temperature of crystallisation of the pure solvent (for example, pure water). This occurs in the nature in the formation of glaciers with huge reserves of crystalline ice formed from fresh water. A similar process of directional freezing-out from the melt is also utilised in the refining of metallic Ga whose melting point is 29.774 °C.

6. CRYSTALLISATION FROM MELTS

Crystallisation from melts is the process of transition of metals and alloys from the liquid L (melt) to the solid state S (crystal). The current theories of the structure of melts include the monoatomic concept in which the structural unit is the atom or the molecule, and the polyatomic concept in which the structural unit is a group of atoms (clusters of atoms). The polyatomic concept can be used, for example, for calculating the coefficient of self-diffusion of elements in the melt, the viscosity of the melt, the change of the volume during the L–S transformation, etc.

The structural components of microinhomogeneous melts are:

- a cluster of atoms, with the atoms in the clusters vibrating;
- the equilibrium zone of broken bonds between the clusters.

This zone is the element of the space and can be characterised by the volume or number of broken bonds. The volume of the zone of broken bonds at the melting point is not large and its value for the majority of metals is 1–6% which approximately corresponds to the experimental data on the change of the volume of metal with the change of state [13]. The broken bonds between the clusters are the reason for the formation of a large number of activated atoms, i.e. the atoms on the surface of the clusters with at least one broken bond. The activated atoms in comparison with the atoms in the volume of the clusters are characterised by high mobility and are transferred into a specific zone. The zone of the activated atoms cannot, however, exist independently because it is part of space and mass. The concentration of the activated atoms in different metals at the melting point (T_m^A) is relatively high, 10–40%. With increase of temperature of the melt, the dimensions of the clusters decrease and this is accompanied by an increase of the number of activated atoms which represents 100% of mass at the evaporation temperature (T_b^A).

From the viewpoint of the polyatomic concept of the structure of the melt, crystallisation or melting can be regarded as a transition of the system from one equilibrium state to another:

Substance (clusters of atoms)	Crystallisation $\rightleftarrows$	Substance (crystal lattice)
Space (breaking of bonds between clusters)	Melting	Space (lattice vacancy)

Generally, the processes of crystallisation or melting are examined in classic thermodynamics. The molar free enthalpy ΔG can be determined thermodynamically by the Gibbs–Helmholtz equation,

$$\Delta G = \Delta H - T\Delta S \tag{17}$$

Phase changes are accompanied by sudden changes of the thermodynamic quantities of molar enthalpy of transformation ΔH and molar entropy ΔS at a constant transformation temperature T which are material parameters for every individual substance, for example, see table P1 in the Appendix. For example, for a change of the liquidus–solidus state (or solidus–liquidus) of compound A, the melting entropy ΔS_m^A can be determined from the relationship:

$$\Delta S_m^A = \frac{\Delta H_m^A}{T_m^A} \qquad [\mathrm{J\ mol^{-1}K^{-1}}] \tag{18}$$

where ΔH_m^A is the molar enthalpy of melting of component A [J mol^{-1}], T_m^A is the melting point of component A [K].

According to Richardson's rule, the majority of metals has the approximate volume $\Delta S_m^A \approx 9.6$ J mol^{-1}K^{-1}, and for typical elementary semiconductor silicon and germanium and other semiconductor compounds $\Delta S_m^A \approx 29$ J mol^{-1}K^{-1}, Figs. 11 and 12. The values of the melting entropy are important for the classification of the crystallising atoms in the formation and stability of the crystal–melt phase boundary (the monoatomic Jackson and polyatomic Temkin concept of build up of the atoms at the crystallisation interface). In the semiconductors, the high value of ΔS_m^A results in the preferential formation of a planar interface, whereas in the metals the interface is 'wrinkled' because of a lower energy barrier (ΔS_m^A). This is also associated with the high concentration of defects in the crystals of metals and alloys.

For the crystallisation or melting processes, the equilibrium conditions at the temperature of the state change are given by the equality of the Gibbs isobaric–isothermal potentials for the melt G_L and the solid phase G_S. At temperatures lower than T_m^A, the solid phase is characterised by a low value of free enthalpy (Gibbs thermodynamic potential) G_S and, consequently, it is characterised by higher thermodynamic stability. At temperatures higher than T_m^A, the liquid phase has higher stability, Fig. 13 [14]. The change of the free enthalpy $\Delta G_V = G_L - G_S$ in the transition of the melt to the solid phase is the driving force that determines the direction of the change of the state.

The previously mentioned basic thermodynamic conditions of the the state are, however, not sufficient, because the actual start of phase transformations is controlled by kinetic factors:

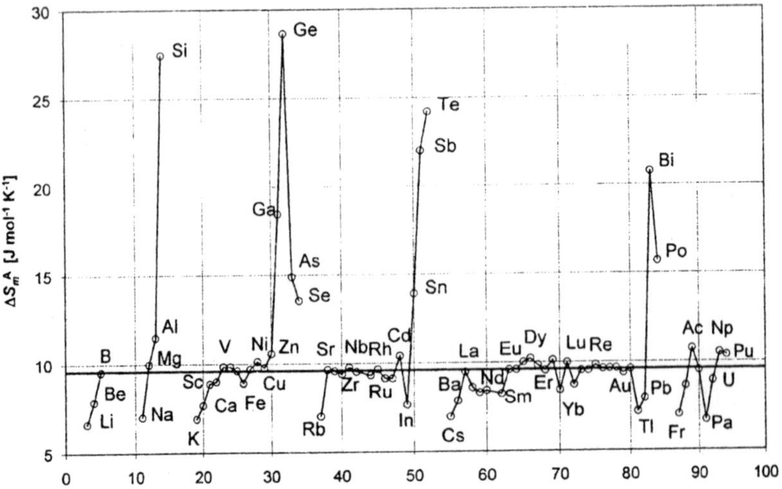

Fig. 11 Entropy of melting as a function of the proton number of the element.

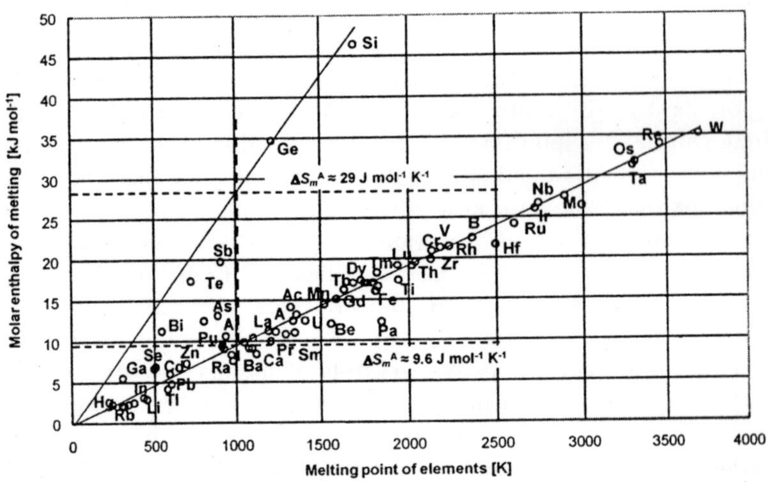

Fig. 12 Enthalpy of melting as a function of the melting point of elements.

- ΔT, temperature supercooling, representing the extent by which the process temperature is lower than the equilibrium liquidus temperature T_L, $\Delta T = T_m^A - T$ for pure compounds, or $\Delta T = T_L - T$ for alloys.
- ΔC, concentration supersaturation which, in the case of alloys, is the deviation of the concentration C in the melt from the equilibrium concentration C_L, i.e. $\Delta C = C_L - C$ (Fig. 14).

The kinetics of the crystallisation processes consists of two independent phenomena:

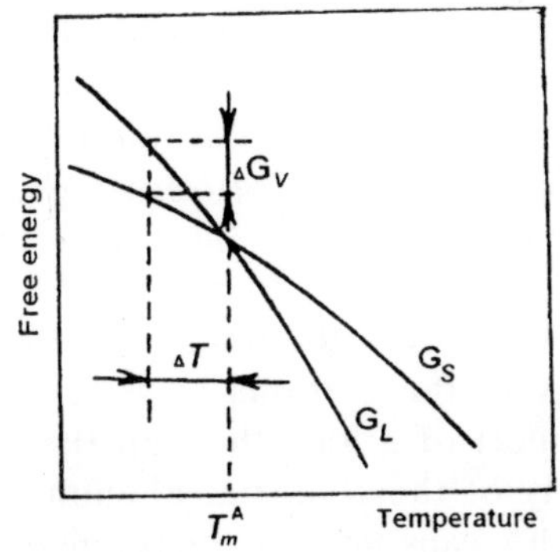

Fig. 13 Dependence of the free volume enthalpy G_V of the solid and liquid phases on temperature.

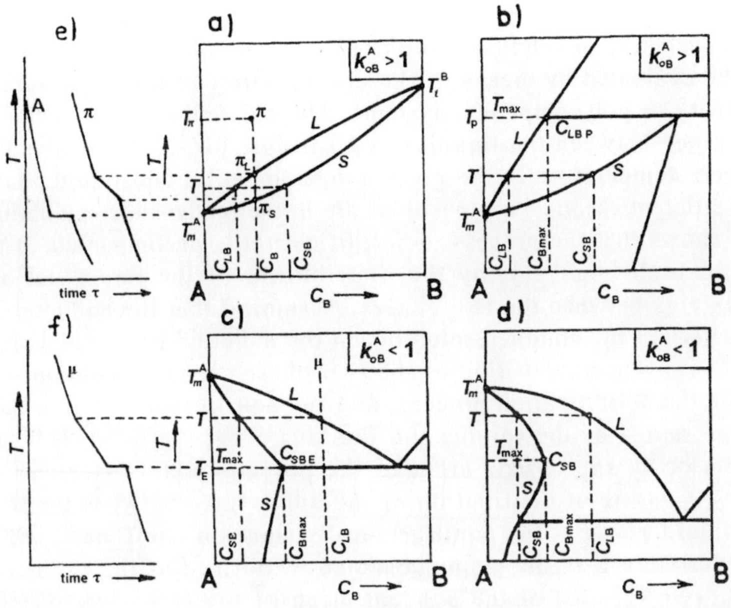

Fig. 14 Main types of binary diagrams and solidification curves. a) unlimited solubility (ideal type), b) peritectic type, c) eutectic type, d) retrograde solubility of the solidus curve, e) curve of cooling the pure substance A and alloy π, f) curve of cooling for alloy μ in the case of the eutectic reaction.

- nucleation, i.e. the formation of crystallisation nuclei (homogeneous or heterogeneous nucleation);
- the growth of crystals, i.e. the rate of displacement of the different types of crystal–melt interface (planar, cellular, dendritic) in relation to the temperature gradient.

Each of these phenomena is a sum of several partial phenomena of which the phenomena with the lowest rate limit the rate of the main process. The nucleation process may include the collection of the present types of atoms by diffusion or other types of motion whose mutual internal arrangement corresponds to the new phase at the solidification front.

The growth of crystals includes the transport of atoms by the 'old' phase (melt), the 'jump' of these atoms to different types of phase boundary and the transport of atoms through the new phase (solid substance). The majority of these partial phenomena are thermally activated processes which means that the energy barriers are overcome by the thermal motion of activated atoms or of groups of these atoms.

6.1. Distribution coefficients in crystallisation

The concentration conditions in the process of solidification of alloys can be evaluated by means of the appropriate equilibrium diagrams of binary or polycomponent systems. The existence of a temperature difference between the liquidus and solidus, Fig. 14, indicates that at every temperature in the given range the solid and liquid phases with different chemical compositions are in thermodynamic equilibrium. This shows that the process of solidification is accompanied, at the crystal–melt boundary, by the distribution of the impurities and admixtures between the two phases. Assuming that the admixture is characterised by unlimited solubility in the melt and in the solid phase (Fig. 14a), the composition of the two phases changes continuously during the solidification process, and the equilibrium between them is represented by the solidus and liquidus curves in the A–B binary system or by appropriate areas of the polycomponent systems.

The measure of distribution of the admixtures between the solid and liquid phase is the equilibrium distribution coefficients k^{A}_{oB} of the admixture B in the main substance A, defined as the isothermal ratio ($T = T_S = T_L$) of the concentration of the admixture element B in the solid phase X_{SB} (solidus) and in the liquid phase X_{LB} (liquidus):

$$k^{A}_{oB} = \frac{X_{SB}}{X_{LB}} \tag{19}$$

where the concentration is given in molar fractions or in atomic %.

The equilibrium distribution coefficient is lower than 1, $k^{A}_{oB}<1$, for systems in which the admixture element B decreases the melting point T^{A}_{m} of the main component A, Fig. 15a, – eutectic systems. For the systems in which the admixture element B increases the melting

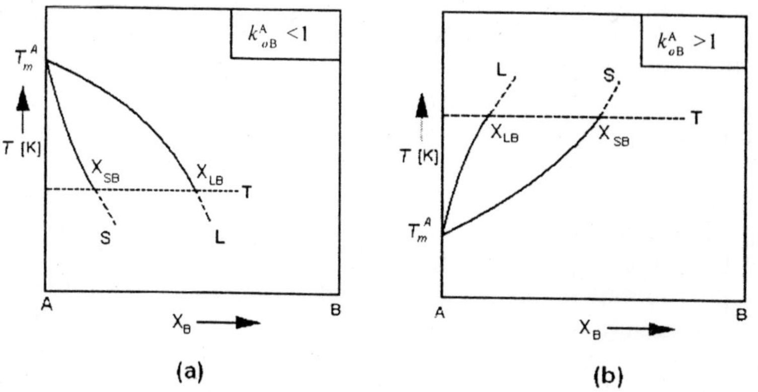

Fig. 15 Definition of the equilibrium distribution coefficient: a) admixture reduces melting point T^A_m (k^A_{oB} <1), b) admixture increases the melting point T^A_m (k^A_{oB} >1).

point, the distribution coefficient is higher than 1, $k^A_{oB} > 1$, Fig. 15b – peritectic types of binary diagrams.

The different solubility of the admixtures and impurities in the melt and the crystal results, because of transfer phenomena, in the distribution of the element.of the matrix (A) and of the admixture (B). This results in the formation, in every primary crystallisation, of different segregation micro- and macroheterogeneities with a negative effect on the properties of solidified substances.

The above-defined distribution coefficient is the main material parameter for the refining of substances by crystallisation processes, such as zone melting and directional solidification, being selective crystallisation processes.

6.1.1. Determination of the equilibrium distribution coefficients from phase diagrams

The main apparatus for the calculation of phase diagrams is thermodynamics. The laws of thermodynamics can be used for the determination of the conditions of equilibrium of both phases and for calculating the form of the curves of the phase diagrams [15–17, Calphad].

The phase data characterising the equilibrium composition of the coexisting phases are usually presented in graphical form as binary diagrams, i.e. a set of the curves $T(X)$ in a functional dependence of the temperature and composition of the phases of the given system [18], or, very seldom, by the numerical values of the coordinates of the points (T,X) for which the system is in equilibrium. In the second case, it is necessary to have suitable mathematical facilities enabling the qualified interpolation of these points, including simultaneous optimisation by means of the criteria of the coexistence of the phase and thermodynamics data [19].

In order to determine the values of the equilibrium distribution coefficients, it is necessary to express, thermodynamically or mathematically, the form of the solidus and liquidus curves. For this purpose, the authors of [6, 7] developed a procedure which can be used for expressing with sufficient accuracy the form of these curves by a polynomial of the second or higher degree, especially in the region adjacent to the main component A in the functional dependences $T_S = f(X_{SB})$, $T_L = f(X_{LB})$ so that they correspond to the actual conditions.

$$T_{S,L} = a_{S,L} X_{S,LB}^2 + b_{S,L} X_{S,LB} + T_m^A \tag{20}$$

For more complicated types of solidus and liquidus curves, to carry out more accurate modelling, it is possible to use, for example, the equation of the cubic parabola (21)

$$T_{S,L} = a_{S,L} X_{S,LB}^3 + b_{S,L} X_{S,LB}^2 + c_{S,L} X_{S,LB} + T_m^A \tag{21}$$

or in the case of the retrograde nature of the solid, which is found especially in semiconductor systems, by the exponential logarithmic function (22):

$$X_{S,LB} = a_{S,L}(\Delta T_{S,L}^{b_{S,L}}) \exp(c_{S,L} \Delta T_{S,L}) \tag{22}$$

The solubility curves of the admixture in the solid solution, for example, in aluminium alloys and others, are efficiently modelled by the exponential function.

The regression parameters of the equations (20)–(22) are determined by the experimental input data from the binary diagrams of the A–B systems on the basis of the original studies by different authors. The form of the curves is controlled thermodynamically, for example, using the criterion proposed by Hayes and Chipman [20,7]:

$$\Delta T_{th} = x_{LB}\left(1 - k_{oB}^A\right) R \frac{\left(T_m^A\right)^2}{\Delta H_m^A} \tag{23}$$

where ΔT_{th} is the thermodynamically calculated deviation of the liquidus temperature from the melting point of the pure component T_m^A at concentration x_{LB}, ΔH_m^A is the molar enthalpy of melting of the pure component A. Equation (23) is valid only in the limiting regions adjacent to the melting point.

The extrapolation of the functional form of the solidus and liquidus curves into the range of diluted solutions ($X_{SB} \to 0$, $X_{LB} \to 0$), used in particular in the preparation of high-purity metals, gives from equation (20) the limiting value of the equilibrium distribution coefficient k^A_{oBlim}

$$k^A_{oBlim} = \frac{b_L}{b_S} \tag{24}$$

The unconventional view regarding the functional form of the solidus and liquidus curves $T_{S,L} = f(X_{S,LB})$ and the functional dependence $k^A_{oB} = f(T, X)$ is clearly indicated by the bilogarithmic form of the binary diagram, Fig. 16. In this representation, the temperature axis gives the value of deviation ΔT from the melting point T^A_m of the pure component ($\Delta T = T^A_m - T$). The horizontal axis gives the range of the concentration of the admixture B up to the limiting region of units of ppm which is especially significant for the metallurgy of high-purity substances with a purity of 5 N or higher. The graph also indicates that the value of k^A_{oBlim} is constant in the majority of cases in the given limiting region from $\Delta T = T^A_m - T < 1$ K, sometimes up to 10 K.

6.1.2. The distribution coefficients of admixtures in copper

A suitable example of the results of mathematical analysis of the solidus and liquidus curves and of the calculations of the distribution coefficients k^A_{oB} are the copper–admixture B binary systems, Fig. 17.

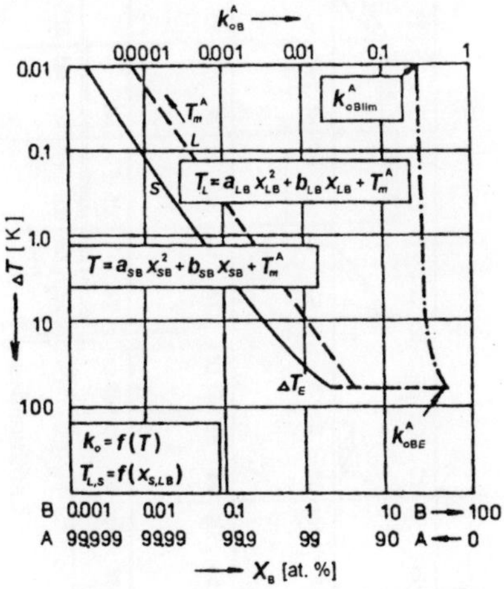

Fig. 16 Bilogarithmic representation of the binary diagram of the A–B system and determination of the value k^A_{oBlim} by graphic extrapolation.

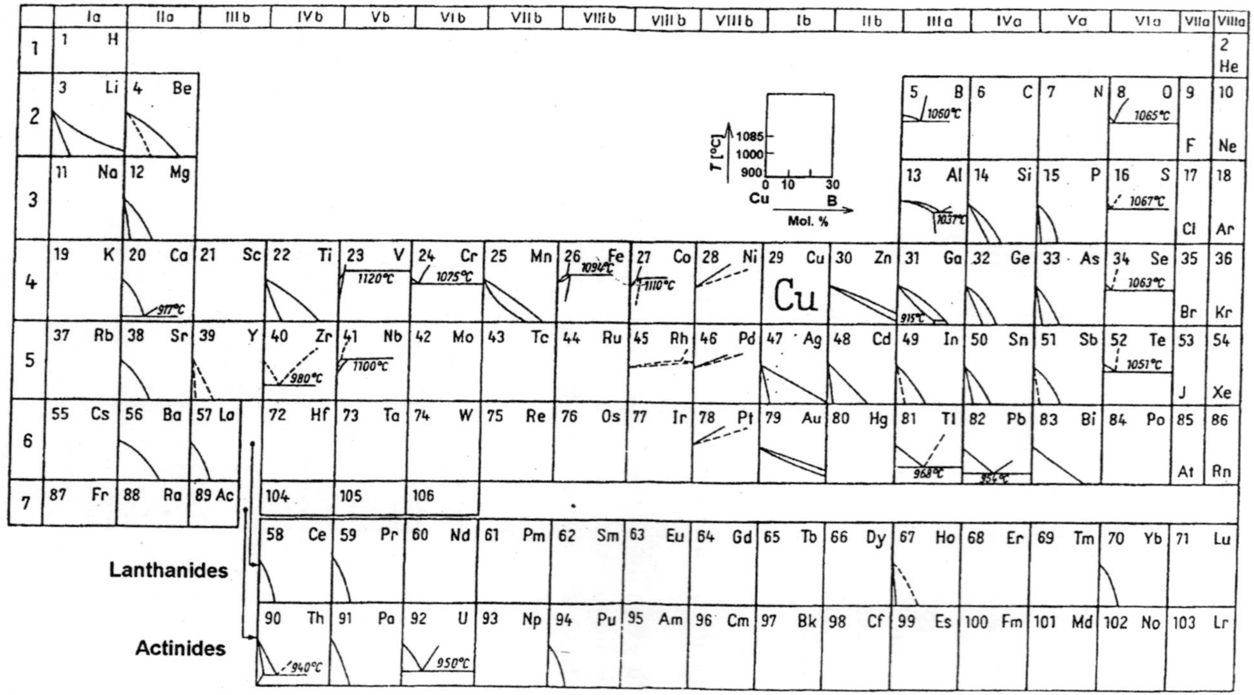

Fig. 17 Solidus and liquidus curves in copper–admixture B binary systems in the region adjacent to the copper side.

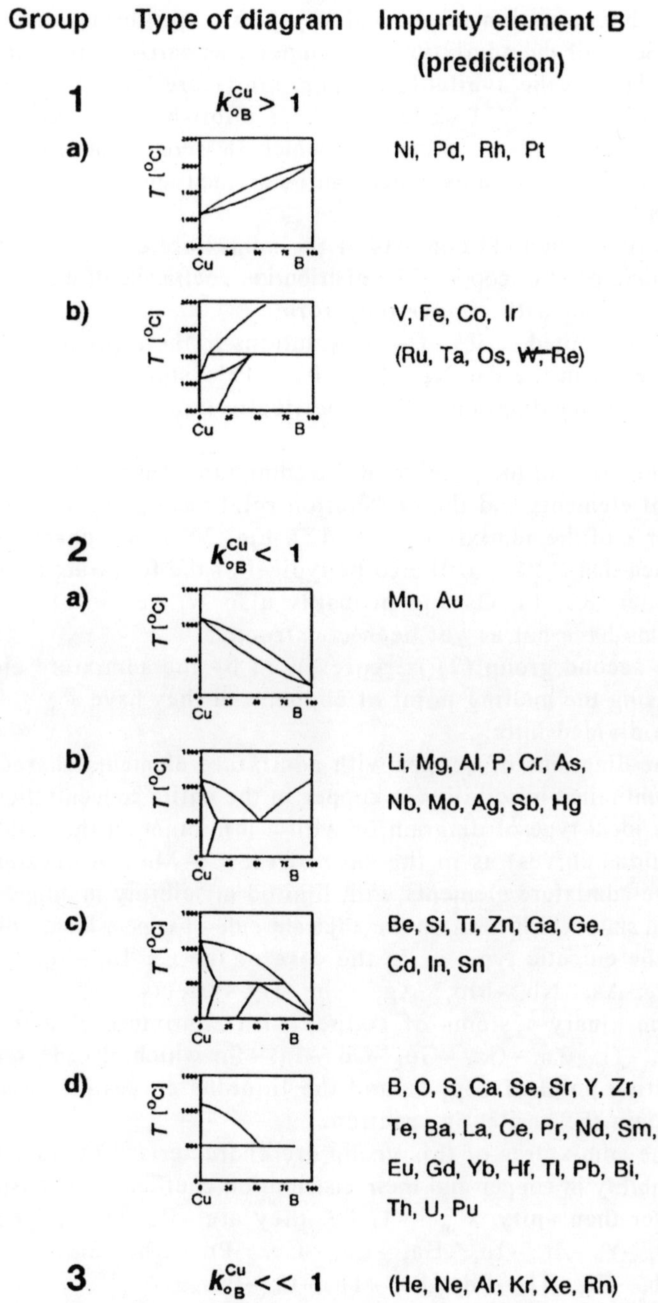

Fig. 18 Types of binary phase diagrams of the copper–admixture system.

The determination of the values of the equilibrium distribution coefficients of the admixtures in copper was carried out on the basis of the data on the available copper–admixture binary diagrams. In reviews [10,21,22,23] various authors published approximately 68 binary copper–admixture systems of which 48 were compiled in complete or partial binary diagrams which can be divided into three main groups, Fig. 18.

The first group (1) consists of the admixture elements increasing the melting point of copper. The distribution coefficient of these elements is $k_{oB}^{Cu} > 1$ and with copper they form:

1a) The unlimited series of solid solutions in the entire concentration range, as in the Cu–Ni, –Pd, –Rh, –Pt systems.

1b) The binary diagrams of the peritectic types Cu–V, –Fe, –Co, –Ir.

On the basis of the position of the admixture elements in the periodic table of elements and the correlation relationship k_{oB}^{Cu} on the proton number Z of the admixture $k_{oB}^{Cu} = f(Z)$, Fig. 20, it was unambiguously predicted that $k_{oB}^{Cu} > 1$ will also be typical of the following admixtures in copper: Ru, Ta, Os and probably also W, Re for which binary diagrams have not as yet been constructed.

The second group (2) is represented by the admixture elements decreasing the melting point of copper and they have $k_{oB}^{Cu} < 1$. They can be divided into:

2a) The diagrams of copper with admixture elements characterised by unlimited miscibility in copper in the entire concentration range (the ideal type of diagram or with a minimum on the solidus and liquidus curves) as in the case of the Cu–Mn, –Au systems.

2b) The admixture elements with limited miscibility in copper in the solid state which form on the adjacent side of copper binary diagrams of the eutectic type, as in the case of the Cu–Li, –Mg, –Al, –P, –Cr, –As, –Nb, –Mo, –Ag, –Sb, –Hg systems.

2c) The binary systems of copper with admixture elements –Be, –Si, –Ti, –Zn, –Ga, –Ge, –Cd, –In, –Sn which also decrease the melting point of copper, and the liquidus curves of the systems contain the peritectic reaction.

2d) The admixtures of this group are characterised by very limited solubility in copper and their distribution coefficient is considerably lower than unity, $k_{oB}^{Cu} \ll 1$, i.e. they are –B, –O, –S, –Ca, –Se, –Sr, –Y, –Zr, –Te, –Ba, –La, –Ce, –Pr, –Nd, –Sm, –Eu, –Gd, –Yb, –Hf, –Tl, –Pb, –Bi, –Th, –U, –Pu.

The groups 2c and 2d include from the position of the elements in the periodic system but, in particular, on the basis of correlation

Table 5 Calculated parameters a_S, b_S, a_L, b_L of equation (20) for Cu–admixture B systems and limiting values of the equilibrium distribution coefficients k^{Cu}_{oBlim} and the values k^{Cu}_{oBEP} for the eutectic or peritectic temperatures; t^{Cu}_m = 1084.87 °C, ΔH^{Cu}_{m} = 13263 J mol^{-1}

Admixture B	k^{Cu}_{oBlim}	k^{Cu}_{oBEP}	a_S	b_S	a_L	b_L	Valid to temperature t [°C]
Ag	0.345	0.081	-8.2434	-22.5990	0.0452	-7.8045	779.1
Al	0.807	0.915	-0.0542	-2.5568	-0.0617	-2.0632	1032
As	0.160	0.377	0.2639	-60.1710	-0.6788	-9.6178	685
Au	0.640	1.00	0.0559	-6.6373	0.0138	-4.2471	889
B	0.06	0.022	-241.0618	-176.6626	0.3878	-10.5650	1013
Be	0.329	0.684	0.6327	-23.7207	-0.0549	-7.8022	866
Bi	0.0001	0.0001	22209967.3	-137998.3660	0.1838	-11.7136	900
Cd	0.275	0.070	-55.2317	-31.7620	-0.0890	-8.7492	549
Ce	0.027	0.049	-158.5567	-421.4238	-1.4097	-11.1908	870
Co	1.436	1.647	-0.0432	3.6593	0.0380	5.2549	1112
Cr	0.546	0.571	0.2437	-9.5326	-0.0770	-5.1996	1074.8
Fe	1.318	1.314	-0.1057	2.8993	-0.1857	3.8214	1096
Ga	0.664	0.768	-0.2918	-5.6058	-0.1998	-3.7209	915
Ge	0.332	0.685	0.0988	-22.9387	-0.4170	-7.6117	824
H	0.119	0.331	-5.0000	-48.5000	-17.5248	-5.7596	1075
Hf	0.053	0.052	-82.0291	-207.5623	-0.1916	-11.0182	988
Hg	0.093	0.331	5.1343	-111.0107	-1.2018	-10.3004	660
Ho	0.0031	0.0021	-353518.1247	-3670.9952	-1.1238	-11.4183	875
In	0.453	0.479	-2.3544	-13.5935	-0.5568	-6.1607	710
Ir	2.863	2.001	-0.1219	7.6129	-2.1295	21.7944	1138
Li	0.270	0.203	1.4223	-31.7660	0.1985	-8.5854	1000
Mg	0.202	0.301	-0.8704	-45.6739	-0.2748	-9.2330	725
Mn	0.449	0.684	0.2383	-14.1723	-0.0797	-6.3653	950
Nb	0.186	0.5	18.6363	-50.5636	-75.0888	-9.3322	1080
Ni	1.478	1	-0.0007	3.7680	-0.0187	5.5698	1455
O	0.019	0.018	-750.4835	-607.7045	0.3031	-11.6330	1066
P	0.091	0.223	2.5604	-114.7884	-0.8414	-10.4141	714
Pb	0.023	0.007	-7932.2808	-480.6312	0.1817	-11.2925	955
Pd	1.083	1	0.0338	0.9481	0.0469	1.0268	1200
Pt	2.026	1	0.0611	5.8596	-0.0550	11.8697	1400
Pu	0.018	0.075	385.7680	-614.9797	-2.2159	-11.0102	855
Rh	2.15	1	-0.0106	6.2506	-0.0231	13.4351	1963
S	0.024	0.016	-13198.3242	-477.0766	-0.5210	-11.3235	1067
Sb	0.088	0.308	7.5739	-119.4973	-0.6664	-10.4910	645
Sc	0.027	0.039	-25.0016	-421.1709	-0.4340	-11.2740	865
Si	0.415	0.698	-0.4531	-15.7790	-0.5004	-6.5501	852
Sn	0.176	0.498	2.1268	-53.5702	-0.5850	-9.4427	798
Sm	0.161	0.50	0.1500	-56.8250	-2.3744	-9.1177	860
Te	0.011	0.002	-240512.4654	-1109.9723	0.8460	-11.7539	1050
Th	0.0007	0.0013	52773.1847	-17248.3894	-1.4328	-11.3848	935
Ti	0.308	0.365	0.1528	-26.1529	-0.0467	-8.0583	885
Tl	0.024	0.019	162.7098	-473.3210	0.2332	-11.4439	968
V	2.46	2	47.2954	6.0762	182.0335	15.0116	1120
Y	0.0072	0.0091	-6098.8293	-1610.7522	-0.7884	-11.5550	871
Yb	0.0015	0.003	3408.7481	-7612.0060	-1.1029	-11.5609	859
Zn	0.723	0.963	-0.0441	-4.3582	-0.0495	-3.1501	902
Zr	0.019	0.014	-2594.3073	-589.6262	-0.1993	-11.4144	972

Table 6 Distribution coefficients k^{Cu}_{oBlim} [Authors, 25] and k^{Cu}_{oB} [26–29]of admixtures in copper

Element B		Authors	[25]	[26]	[27]	[28]	[29]	Element B		Authors	[25]	[26]	[27]	[28]	[29]
H	1	0.12		0.6				Mo	42			3			
He	2	<0.001						Rh	45	2.15	2.13	> 1			
Li	3	0.27	0.3					Pd	46	1.09	1.07	2	7.15		0.65
Be	4	0.33	0.34	0.35	0.06			Ag	47	0.34	0.3	0.4	0.58	0.45	0.13
B	5	0.06	0.052	0.05	0.04			Cd	48	0.27	0.064	0.12	0.18		
C	6		~ 0.1	0.1				In	49	0.45	0.4	0.45	0.57		0.12
O	8	0.019	0.022	0.01				Sn	50	0.18	0.16	0.12	0.15		
Ne	10	<0.001						Sb	51	0.088	0.17	0.1	0.14		
Mg	12	0.2	0.11	0.15	0.09			Te	52	0.011					
Al	13	0.81	0.93	0.9	~ 1			Xe	54	<0.001					
Si	14	0.42	0.43	0.5	0.26	0.38		Ba	56			0.01			
P	15	0.091	0.11	0.2	0.1			La	57			(0.06)			
S	16	0.024	0.017	0.01				Ce	58	0.027	0.024	(0.07)			
Ar	18	<0.001						Pr	59			(0.08)			
Ca	20		0.05	0.05				Nd	60			(0.07)			
Sc	21	0.027		0.1				Pm	61			(0.07)			
Ti	22	0.31	0.27	0.1	0.6			Sm	62	0.16	0.094	(0.07)			
V	23	2.46		1.1				Ho	67	0.003					
Cr	24	0.55	0.54	0.4	0.5	0.5		Yb	70	0.0015					
Mn	25	0.45	0.37	0.6	0.52	0.7		Hf	72	0.053		(0.6)			
Fe	26	1.32	1.32	1.4	1.77	2	1.3	Ta	73			(8)			
Co	27	1.44	1.53	1.2	1.82		1.1	W	74			(3.3)			
Ni	28	1.48	1.68	3.4	7.1	2.6	4.2	Re	75			(3)			
Cu	29	**1.00**	**1.00**					Ir	77	2.86	2.42				
Zn	30	0.72	0.72	0.7	0.67			Pt	78	2.03	1.79	2	1.68		
Ga	31	0.66	0.59	0.5	0.57			Au	79	0.64	0.5	0.5	0.69		
Ge	32	0.33	0.4	0.5	0.32			Hg	80	0.093	0.11	(0.2)			
As	33	0.16	0.19	0.2	0.21		0.17	Tl	81	0.024	0.072	0.2			
Se	34			0.006				Pb	82	0.023		< 0.01			
Br	35			0.002				Bi	83	0.0001		<0.001			
Kr	36	<0.001						Rn	86	<0.001					
Sr	38			0.025				Ac	89			(0.025)			
Y	39	0.0072	0.056	0.1				Th	90	0.0007	0.084	0.025			
Zr	40	0.019	0.02	0.1				U	92			(0.025)			
Nb	41	0.18	0.21	8				Pu	94	0.018	0.01				

periodicity (Fig. 20), also other admixture elements in copper for which binary diagrams are not yet available, as is the case with other lanthanides, alkaline metals, halogens and actinides.

The third group is represented by inert gases –He, –Ne, –Ar, –Kr, –Xe, –Rn, which have almost no solubility in copper and the distribution coefficients of these elements in copper are considerably lower than unity ($k^{Cu}_{oB} \ll 0.001$).

Generally, it may be concluded that as the distance of k^{Cu}_{oB} from unity increases, i.e. $k^{Cu}_{oB} \gg 1$ and $k^{Cu}_{oB} \ll 1$, the efficiency of refining of copper to remove the given admixture in solidification refining processes, such as zone melting and directional solidification, is higher.

Table 5 presents the calculated regression parameters a_S, b_S, a_L,

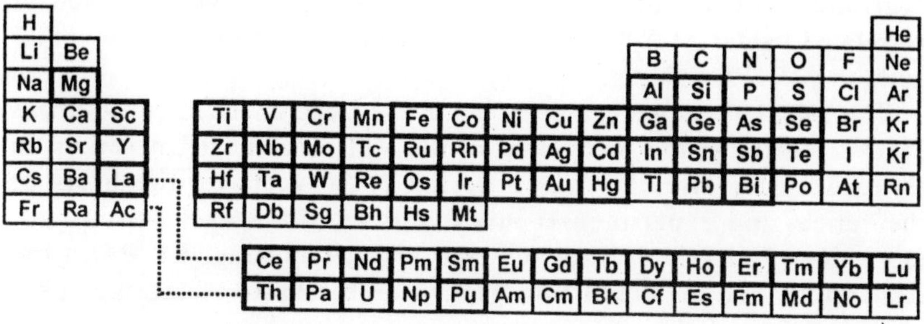

Fig. 19 Examined binary diagrams of main elements.

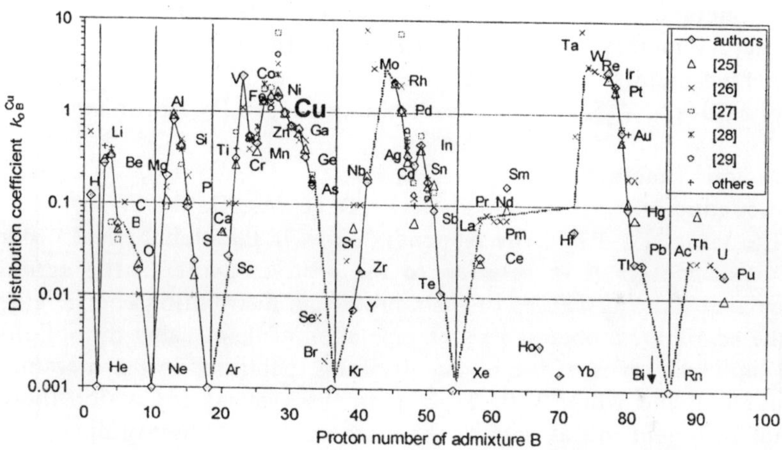

Fig. 20 Periodic correlation dependence of the distribution coefficients k_{oB}^{Cu} of the admixtures in copper on the proton number of the admixture.

b_L of the equations of the type of (20) of the solidus and liquidus for the individual admixtures in copper from the data base prepared by the authors of the present book [24], including the temperature range of validity (from the melting point of copper t_m^{Cu} = 1084.87 °C up to the given temperature). Using the curves plotted in this manner, it was possible to determine the limiting values of the equilibrium distribution coefficients of the admixtures in copper $k_{o\lim}^{Cu}$ and the values of the distribution coefficients k_{oBEP}^{Cu} valid for the eutectic or peritectic temperature, Table 5.

Table 6 presents, in dependence on the increasing proton number of the admixture, the limiting values of the equilibrium distribution coefficients of the admixtures in copper k^{Cu}_{oBlim} and the predicted or correlated values of k^{Cu}_{oB}, including the values of k^{Cu}_{oB} obtained by other authors [25–29, 6,7].

The calculations were carried out at the Department of Non-ferrous Metals, Refining and Recycling of the VSB-Technical University in Ostrava systematically over a period of 30 years on the basis of theoretical and experimental investigations of the A–B systems for more than 50 basic substances (see Fig. 19 and Fig. P1–P7 in the Appendix), and arranged into groups on the basis of the technical classification of the elements:

Iron	Fe
General non-ferrous metals	Co,Ni,Cu,Ga,Cd,Zn,Hg,In,Sn,Sb,Pb
Light metals	Mg, Al
Noble metals	Ag,Au,Ru,Rh,Pd,Os,Ir,Pt
Refractory metals	Ti,Zr,Hf,V,Nb,Ta,Cr,Mo,W,Re
Dispersed metals	Sc,Y,La
Lanthanides	Ce,Pr,Nd,Sm,Gd,Tb,Dy,Ho,Er,Tm, Yb,Lu
Radioactive metals	U, Th, Pu
Semiconductors	Si,Ge,Se,Te (GaAs, GaSb, CdTe)

The tables P2–P7 in the Appendix show in the alphabetical order the basic metals A in relation to the proton number Z the actual calculated limiting values of the equilibrium distribution coefficients of the admixtures, obtained by extrapolation methods using the solidus and liquidus curves of the binary diagrams, published in the literature throughout the world [10, 21–23]. In the systems for which there is not sufficient initial data or there are no data for binary diagrams, correlations were used to process the prognosis of the values by means of the signs <1 or >1 or ≈ 1. This also relates to the values for the inert gases helium, neon, argon, krypton, xenon and radon (<0.001).

6.1.3. Correlation dependence of distribution coefficients

Another significant function of the equilibrium distribution coefficients is the inclusion of these coefficients in different correlation relationships for a number of physical properties of the admixture elements, such as, for example, melting point, standard sublimation energy, tetrahedral, covalent or ion radii, the lattice energy, the valency of the atoms or the maximum solubility of the admixtures in the solid, etc. However, the most important function is the periodic correlation dependence of the distribution coefficients of the admixtures in the main substances on the proton number of the admixtures. The distribution coefficient represents an entire complex of the properties of the admixture element manifested during its implantation in the lattice of the main substance.

A suitable example of the correlation dependence is the periodic correlation dependence of the values of the distribution coefficients of the admixtures in copper on the proton number of the admixtures, Fig. 20, which consists of the values of k^{Cu}_{oB} (from Table 6). In this graphical relationship, the minima of the curves are represented by the values of k^{Cu}_{oB} of the inert gases He, Ne, Ar, Kr, Xe and Rn, which are separated by the individual periods from each other. In the first three short periods, maxima are found for the values of k^{Cu}_{oB} of the admixtures H, Li and Al. In large periods, the maxima are evident in the elements V, Co, Ni (period 4), Mo, Rh (period 5), and Ta, W, Re, Ir (period 6). For the lanthanides, the number of binary diagrams with copper available at present is very small. However, on the basis of the correlation of k^{Cu}_{oB} with the proton number it may be expected that their values of k^{Cu}_{oB} will be smaller than 0.1.

Similar periodic correlations were already plotted previously and verified for more than 50 basic elements [6,7,30,31], see the Appendix (Fig. P8–P17). The priority of the discovery of periodicity k^{Al}_{oB} on the proton number of the admixtures in aluminium (Fig. P8) [32, 33] was granted by V.N. Vigdorovich [34] to Kuchar already in 1966.

Researchers at the VSB – Technical University of Ostrava have developed a database of the investigated binary systems containing the limiting values of the equilibrium distribution coefficients k^{A}_{oBlim}, the values of the distribution coefficients for eutectic k^{A}_{oBE} or peritectic reactions k^{A}_{oBP}, the parameters of the curves of the solidus a_S, b_S and liquidus a_L, b_L (20) and the temperature range of validity of both curves from t^{A}_{m} to T. This database [24] contains at present information on more than 1300 binary systems and the authors of the book have the information available for experts working in the area of refining substances by crystallisation methods, for metallurgists working with liquid phases and for casting technology.

The periodic correlation dependences of the equilibrium distribution coefficients of the admixtures in basic metals on the proton number of the admixtures generally make it possible to:

- determine the unknown values of k^{A}_{oB} and predict the distribution of the admixtures during both selective and general crystallisation processes;
- indicate the suitability of practical realisation of zone melting or directional crystallisation for the preparation of high purity substances;
- select suitable initial materials for these processes;

- carry out preliminary evaluation of the attainable degree of refining;
- carry out objective control of crystallisation processes in refining metallurgy;
- carry out controlled microalloying and doping of the admixtures during the growth of single crystals and commercial alloys thus increasing the physical–metallurgical properties;
- calculate the concentration supercooling in solidified materials at the crystal–melt boundary and, consequently, predict the growth structure;
- predict the main types of binary diagrams;
- predict the distribution capacity and enrichment of foreign admixtures with $k_{oB}^{A} > 1$ in dendrite arms, build up of admixtures with $k_{oB}^{A} < 1$ in the spaces between the dendrites, in the mother liquid during dendritic segregation which almost always accompanies the actual solidification of substances;
- calculate the decrease or increase of the melting point of the main component at the given concentration of the admixture;
- determine the width of the solidification range which must be known for controlling the production processes of commercial alloys in conventional or continuous casting and solidification of substances.

The distribution coefficients are the main material parameter of solidifying substances. All theoretical and experimental results have been applied not only in refining processes but also in other material regions if they utilise the crystallisation processes taking place at the crystal–melt interface.

6.2. Interphase (kinetic) distribution coefficient

In the real crystallisation process, there is no thermodynamic equilibrium of the individual phases at the crystal–melt interface. The phase transformation takes place at a specific rate so that the liquid and solid phases are characterised by the presence of concentration gradients, and the temperature at the interface is somewhat lower than the equilibrium temperature.

The phase boundary is characterised by the occurrence of two processes, i.e. the actual phase transformation and the transport processes of the participating phases (mass and energy transfer). In the solidification of pure metals and the majority of other substances, the rate of crystallisation determines the removal of latent heat from the interface. The admixtures with $k_{oB}^{A} < 1$, which are implanted in the crystal during solidification with a lower concentration in comparison with the liquid phase, enrich the melt in front of the phase boundary

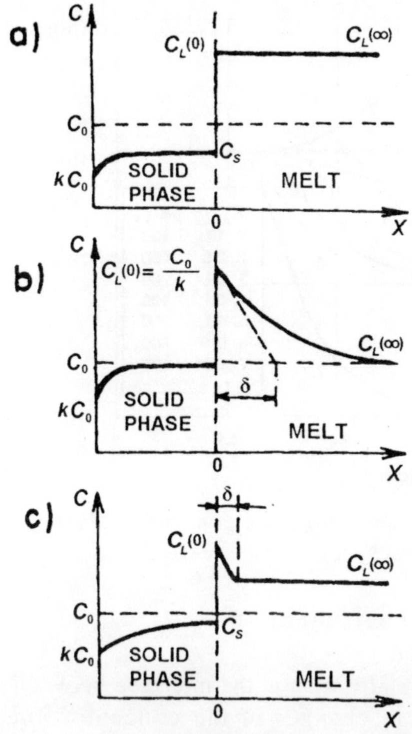

Fig. 21 Distribution of an admixture in the vicinity of the phase boundary ($k^A_{oB} < 1$).

and must be transported into the main volume of the liquid phase, Fig. 21. The behaviour of the admixture element B at the crystal–melt phase boundary is described by the so-called interphase or kinetic distribution coefficient k^{A*}_{B}:

$$k^{A*}_{B} = \frac{x_{SB(0)}}{x_{LB(0)}} \tag{25}$$

In many cases, the distance of the interface with its immediate environment from the equilibrium is so small that the interphase distribution coefficient is approximated with sufficient accuracy by the equilibrium distribution coefficient [6,7]:

$$k^{A*}_{B} \cong k^{A}_{oB} \tag{26}$$

However, the kinetic distribution coefficient also depends on the morphology of the atomically smooth or atomically rough (diffusion interface) phase boundary and on the crystallographic orientation of the growth areas, facets and growth nuclei. In this case, k^{A*}_{B} is not equal to k^{A}_{oB}.

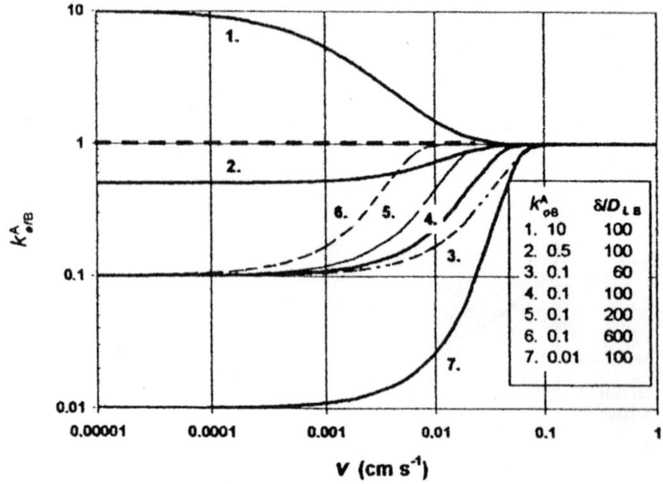

Fig. 22 Dependence of the effective distribution coefficient k^{A}_{efB} on the rate of growth and on the δ/D_{LB} ratio for different values of k^{A}_{oB}.

6.3. The effective distribution coefficient

The processes taking place during crystallisation at the interface between the liquid and solid phases result in changes of the concentration in the melt in the immediate vicinity of the interface, Fig. 21. The concentration of the admixture in the melt is equalised by the effect of different types of convection in the main volume of the melt and by diffusion in the limiting layer with thickness δ. The relationship for the effective distribution coefficient of the admixture in the main substance was derived by Burton, Prim and Slichter [35] from the equation of continuity and for suitable boundary conditions:

$$k^{A}_{ef\,B} = \frac{k^{A}_{oB}}{k^{A}_{oB} + \left(1 - k^{A}_{oB}\right)\exp\left(\dfrac{-v\,\delta\,\rho_S}{D_{LB}\,\rho_L}\right)} \tag{27}$$

where v is the macroscopic speed of movement of the phase boundary [cm s^{-1}], δ is the thickness of the sublaminar diffusion layer [cm], D_{LB} is the diffusion coefficient of the admixture B in the melt [cm^2 s^{-1}], ρ_S, ρ_L are the densities of the solid and liquid substances [kg m^{-3}].

The theory proposed by Burton, Prim and Slichter assumes that the solidification front is characterised by the formation of a thin diffusion layer δ whose thickness changes in relation to the intensity

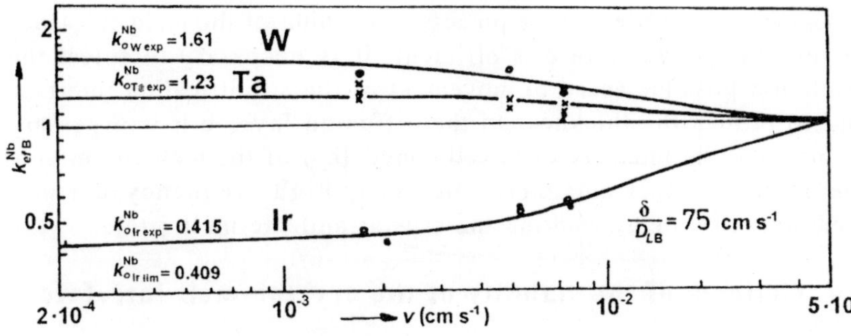

Fig. 23 Comparison of the experimentally determined values of the distribution coefficients k^{Nb}_{oBexp} of the admixtures W, Ta and Ir in niobium after electron beam zone melting with the values of k^{Nb}_{oBlim}.

of mixing the melt. The thickness of this layer also depends on natural convection and viscosity of the melt, and the speed of rotation of the crystal (forced convection) and on the nature of the flow in the melt (turbulent or laminar) and only to a certain degree on the crystallisation rate.

The values of the diffusion coefficient of the admixture element in the melt D_{LB} are often not available. For many melts, these values are in the range between 10^{-3} and 10^{-5} cm^2 s^{-1} in comparison with the considerably lower value of the diffusion coefficients in the solid phase D_{SB} which at temperatures slightly below the solidus is usually three orders of magnitude lower, 10^{-6} to 10^{-10} cm^2 s^{-1}.

Equation (27) indicates that at the growth rate $v = 0$ the value $k^{A}_{efB} = k^{A}_{oB}$ which means that as the solidification rate decreases, the degree of enrichment by the impurity in the melt for the case in which $k^{A}_{B} < 1$ increases. At the rate $v = \infty$, $k^{A}_{efB} = 1$, and there is no distribution of the admixture in the macrospace of the ingot, Fig. 22.

The effective distribution coefficient can be determined by experiments and is described by the following equation (Fig. 21):

$$k^{A}_{ef\,B} = \frac{x_{SB(O)}}{x_{LB(\infty)}}. \qquad (28)$$

If several experimentally determined values of k^{A}_{efB} for different speeds of movement of the solidification front v are unavailable (maintaining the constant values of other parameters in such a manner that δ/D_{LB} = const), by extrapolating the rate $v \rightarrow 0$ we can obtain the value of the distribution coefficient $k^{A}_{o\,Bexp}$, Fig. 23 and Table P7 in the Appendix [159].

In order to ensure that the process of refining of the main substance to remove the admixture is efficient, it is necessary to select the minimum possible speed of movement of the solidification interface and to reduce the thickness of the diffusion layer δ it is necessary to increase the intensity of forced convection of the melt (by mixing the melt by electromagnetic methods, high frequency current, mechanically, or by rotating the crystal and the melt, etc).

6.4. Criteria of the stability of the crystal–melt interface

In addition to the actual transformation, represented by nucleation, the liquid–solid phase boundary is also characterised by the occurrence of transport processes in the phases contributing to the phase transformation. Both the actual mass of the solidifying substance with mass and concentration changes at of the liquid–solid interface, and also the energy (melting enthalpy) which is released during crystallisation at the interface, are transported.

a) Temperature conditions of crystallisation

For the continuous growth of the crystal it is necessary to ensure

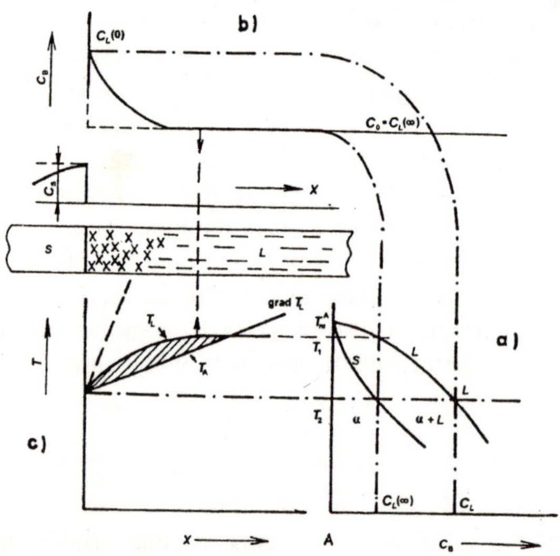

Fig. 24 Conditions of formation of concentration supercooling: a) part of the binary diagram of the A–B system, b) concentration profile of component B at solidification interface, c) the distribution of the equilibrium isoliquidus temperature T_L and real liquidus temperature T_A at the solidification interface.

the constant removal of the latent heat of solidification (solidification enthalpy) from the phase boundary. However, this is only possible in the presence of a specific temperature gradient in the region at the centre of the solid–liquid interface at which the movement of the interface starts. During solidification of pure substances the rate of crystallisation of the transformation is generally given by the removal of heat from the interface. At the melt–solid interface, two different types of heat removal may form from the viewpoint having a controlling effect on the crystallisation process:

- the heat is removed only by the solid phase – the temperature gradient in both the solid and liquid phases is positive. In the case of a low growth rate, the interface is flat and is characterised by the occurrence of directional crystallisation.
- the melt is supercooled, the transfer of heat takes place through the solid phase and the melt, the temperature gradient in the melt is negative, the growth rate is high and at the interface the dendritic growth of the crystals takes place.

The temperature gradient grad T_L [K cm^{-1}] is positive if the temperature in front of the solidification plane is higher than at the phase boundary. If the temperature gradient is negative, the temperature in front of the solidification plane is lower than the solidification temperature and the melt is supercooled.

b) Concentration supercooling in crystallisation

The examination of the A–B binary system with the distribution coefficient $k^{A}_{oB} < 1$, Fig. 24a, shows that in the system the liquidus temperature decreases with increasing concentration of the component B. The admixture B with $k^{A}_{oB} < 1$ is enriched in the melt ahead of the solid–liquid crystallisation interface, Fig. 24b. The concentration of the admixture B changes, according to theoretical calculations, with the distance x from the interface in an exponential manner and every point in the melt ahead of the solidification interface has the known concentration of the admixture B. This is characterised by the formation of an approximately exponential increase of temperature dT_L/dx at the distance x ahead of the solidification front, because every corresponding point of this so-called isoliquidus T_L also has the defined equilibrium temperature, according to the appropriate binary diagram, curve T_L in Fig 24c. Cooling of the molten metal results in the formation, in the melt of the temperature gradient grad $T_L = dT_A/dx$, indicating the increasing linear course of the actual temperature T_A, Fig. 24c, which increases in the melt in front of the interface. Because of this positive temperature gradient, temperature T_A in certain areas ahead of the solidification interface is lower than the temperature of the isoliquidus T_L, Fig. 24c.

According to B. Chalmers, this phenomenon is referred to as the concentration supercooling and in Fig. 24c it is indicated by the cross-hatched region between the curve of the isoliquidus T_L and the actual temperature in the melt T_A represented by the temperature gradient grad T_L. This region is characterised by the concentration supercooling. In the presence of a steep temperature gradient and a low degree of enrichment of the admixture B in front of the phase boundary and in the case of low solidification rate v, the actual temperature T_A in the melt is higher than the isoliquidus temperature T_L, i.e., in the limiting case, T_A represents the tangent to the curve T_L at the point $x = 0$. Only in this case can the phase boundary be regarded as planar which, according to Tiller, is the limiting condition for the stability of the plane the interface when there is no concentration supercooling in the given system. This condition of the stability of the planar interface can be expressed by the following relationship [36]:

$$\frac{\text{grad } T_L}{v} \leq \frac{m_{LB}\, C_0 \left(1 - k_{\mathrm{B}}^{\mathrm{A}}\right)}{D_{LB}\, k_{\mathrm{B}}^{\mathrm{A}}} \tag{29}$$

where grad T_L is the gradient of temperature at the solidification front in the direction into the melt [K m^{-1}], v is the speed of movement of the solidification front [ms^{-1}]; m_{LB} is the tangent to the liquidus curve at the concentration C_0; D_{LB} is the diffusion coefficient of the admixture in the melt; C_0 is the initial concentration of the admixture in the alloy; $k_{\mathrm{B}}^{\mathrm{A}}$ is the distribution coefficient.

The movement of the surface of the interface between the solidus and the liquidus phase, i.e. the adsorption of the atoms on the matrix of the metallic atom with the regular structure takes place by the diffusion transfer of heat and in the case of alloys is also by mass transfer. Consequently, the flat (planar) interface remains stable only up to a specific critical growth rate and temperature gradient. The maintenance of the equilibrium condition on the moving solidification front results in the destabilisation of the growth surface in the solidifying melt leading to the corresponding increase of the growth rate or a decrease of the temperature gradient.

The tendency for a decrease in the degree of supersaturation and the associated concentration supercooling in the vicinity of the solidification front increases the surface of the crystal and results in the loss of stability. Consequently, the solidification interface is characterised by the formation of a cellular or dendritic structure.

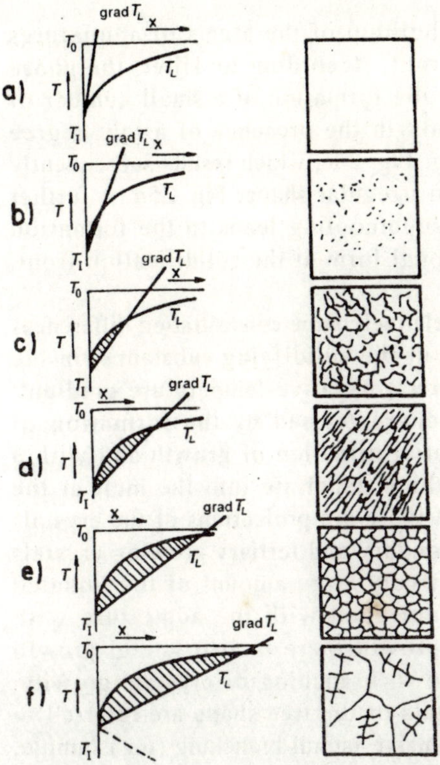

Fig. 25 Relationship of the degree of concentration supercooling with the morphology of growth structures of crystals.

The concentration supercooling of the melt is of primary importance for the formation of solidification nuclei and the atomic kinetics and growth anisotropy. It also affects the morphology of the interface, the size of the crystals and primary grains and the distribution of defects in the crystal. The formation of regions with the supercooled melt leads to the instability of solidification and in real alloys it is a condition for the formation of branched structures of the cellular and dendritic type, Fig. 25.

- the flat (smooth) surface of the interface may form only during crystallisation of very pure metals and isotropic atomic kinetics of growth, Fig. 25a;
- the layer (stepped) surface formed during the anisotropic atomic kinetics of growth observed in metals with the different deposition of the atoms in different crystallographic directions or in the presence of foreign atoms with the preferential absorption on specific crystal planes, Fig. 25b;
- the cellular surface of the interface is characteristic of concentrational

supercooling caused by the buildup of the atoms of admixtures ahead of the solidification front. According to Tiller, the phase boundary is characterised by the formation of a small number of the regular projections already in the presence of a low degree of concentrational supercooling, Fig. 25c, which results subsequently in the formation of cells of an irregular shape, Fig. 25d. A further increase of concentrational supercooling leads to the formation of cells of the regular hexagonal form at the solidification front, Fig. 25e.

the dendritic surface forms at relatively large concentration differences in the content of admixtures in the solidifying substance, in the presence of a very flat or even a negative temperature gradient. The solidification front is characterised by the formation of projections with the preferential direction of growth and with a small radius of curvature, which penetrate into the melt at the higher rate, Fig. 25f. Between these are projections of the crystals characterised by primary, secondary and tertiary growing crystals into the melt, there is a relatively large amount of the liquated mother liquor (segregate), enriched with the admixture with $k^A_B < 1$. In addition to front growth there is also lateral growth accompanied by the well-known branching of crystals growing through the melt. These crystals of the tree shape are referred to as dendrites. If there is no extensive lateral branching (for example, in the case of preferential heat removal), long-axis columnar crystals form on the microscale, for example, in the solidification of the melt in the ingot mould at the interface growing in the direction from the cooling wall of the mould into the melt.

The result of cellular and, in particular, dendritic growth is the formation of inhomogeneities in the chemical composition known in metallurgy as dendritic microsegregation. At the nodal points of the hexagonal cells, as in the case of inter-dendritic spaces, there is intensive concentration of the admixture with the distribution coefficients $k^A_B < 1$. The admixtures with $k^A_B > 1$ result in preferential enrichment of the axis of the dendrites (or cells).

6.5. Directional crystallisation

At the present time, the processes of crystallisation from the melt

---

*In the following text, the distribution coefficient wili not be specified in detailed by means of the arguments $^A{}_B$, because it is always the distribution of the admixture B in the basic substance A.

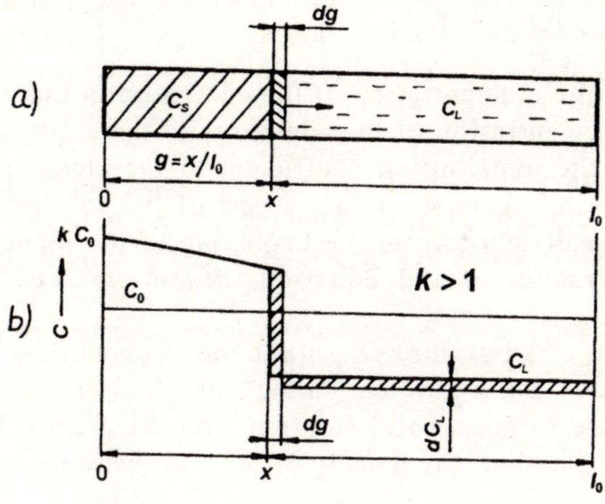

Fig. 26 Principle of directional crystallisation.

[5–7] occupy one of the most important areas in the group of the methods of refining metals and semiconductors. They are used in the final stage of refining for the removal of very small and trace amount of admixtures.

The refining methods, based on the processes of crystallisation from the melt, include:

- directional crystallisation
- zone melting
- drawing of crystals and single crystals from the melt

Directional crystallisation is a refining method based on the distribution of admixtures between the solid and liquid phases. The method is based on the fact that the crystal–melt phase boundary moves slowly through the molten ingot of the refining material with the length l_0 at the initial concentration of the admixture C_0 (Fig. 26). In this case, the solidifying interface is characterised by the distribution of impurities and admixtures present in the basic substance.

The concentration profile of the admixture in the solidified part $C_1(x)$ depends on the change of the concentration conditions at the phase boundary which are given mainly by the values of the distribution coefficients of the individual admixtures present in the main refining substance.

The relationship between the distribution coefficient, the amount of the solidifying substance and the concentration of the admixture was published by the W.G. Pfann [5] (originally by E. Scheil) for directional crystallisation in the following form:

$$C_1(x) = C_0k\,(1-g)^{k-1} \tag{30}$$

where $C_1(x)$ is the concentration of the admixture in the crystal; C_0 is the initial concentration of the admixture in the entire volume of the melt; k is the distribution coefficient of the element B in the main substance A*, g is the relative part of the solidified crystal ($g = x/l_0$), x the distance from the beginning of the ingot.

In the derivation of the equation, Pfann use the following assumptions:

- the diffusion of admixture D_{LB} in the melt is complete and ideal
- the diffusion of admixture D_{SB} in the solid phase is equal to zero
- the admixture forms a solid solution with the main substance,
- during solidification, the density of the material does not change ($\rho_S/\rho_L = 1$)

At $x = 0$, the newly formed concentration is $C_1(0) = C_0k$. As the distance of the value of the distribution coefficient from 1 increases ($k \ll 1$, $k \gg 1$), the distribution efficiency of the given admixture in the entire volume of the ingot increases.

In refining by directional crystallisation special attention is given to important admixtures whose distribution coefficient is lower than 1. In this case, the movement of the solidification boundary through the solidifying crystal is accompanied by gradual enrichment of the melt with the admixture with $k < 1$, i.e. in the enrichment of the end part of the solidified ingot with the total length l_0. On the other hand, the beginning of the ingot is depleted in the admixture with $k < 1$ (Fig. 27). The situation is reversed in the case of admixtures with $k > 1$ which enrich the initial parts of the solidifying ingot and the end part of the ingot is depleted in these admixtures.

However, conventional directional solidification is a one-off process. If the ingot with the resultant concentration profile $C_1(x)$ is remelted, this results, due to convection flows in the melt, in the equalisation of the concentration of the admixture in the entire volume of the melt to the initial value C_0. After repeated directional crystallisation, the identical concentration profile $C_1(x)$ forms along the entire length of the ingot.

In order to ensure efficient directional solidification, the contaminated part of the ingot must be removed. In practice, this is carried out by two methods:

a) the first method is based on the sectioning of the solidified ingot, i.e. the removal of the contaminated part. This is followed by the assembling of cleaner sections and several ingots, with subsequent repeated remelting by directional solidification.

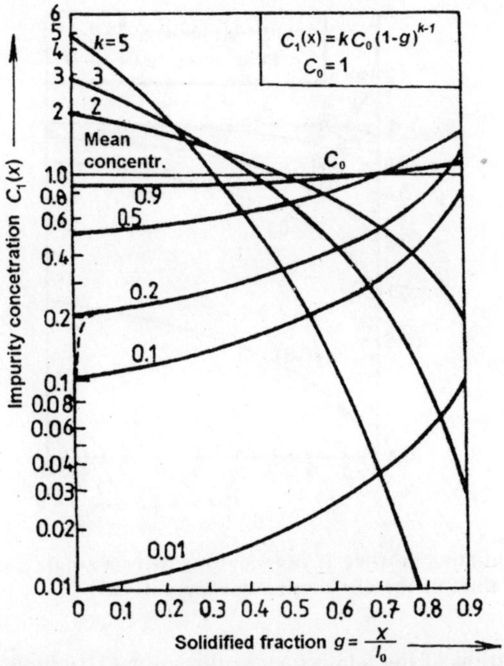

Fig. 27 Concentration profiles of directional crystallisation for different values of the distribution coefficient *k*.

b) the second method is based on melting the material in a boat of a special design. This design enables the removal of the enriched fraction λ of the mother melt from the end part of the ingot, Fig. 31, which would be last to solidify and which is expected to have a higher concentration of the admixtures with $k < 1$. After the removal of the enriched portion λ, the initial level C_0 of the admixture is reduced in the boat to C_p. If this process is repeated several times, it is the multiple directional solidification with the removal of material [7,37].

6.6. Zone melting

Zone melting is a variant of directional solidification used for the refining of substances to obtain high purity and takes place during the controlled distribution of impurities and admixtures present in the ingot by the periodic repetition of the refining crystallisation process in the restricted part, the zone, of the melted ingot, Fig. 29a.

In zone melting, in an ingot with length l_0 only a defined part is melted, i.e. a narrow zone, with the width *b*. The molten zone with

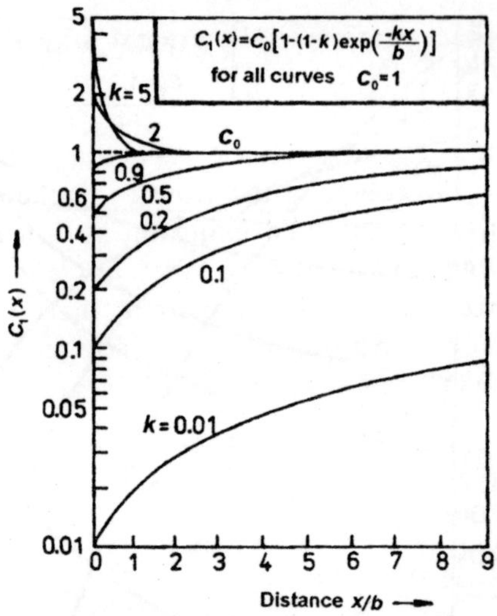

Fig. 28 Concentration profiles of zone melting after a single passage of the zone for different values of the distribution coefficients k.

the concentration of the admixture in the melt C_L, which travels through the ingot, is characterised by two interfaces between the melt and the solid phase:

- the melting front $(x+b)$, in which the initial solid phase of the concentration C_0 is melted in the zone;
- the solidification front, in the area (x), where the material with concentration C_L melted in the zone and homogenised by convection in the melt, is again solidified, with a new concentration $C_1(x)$.

The concentration profiles of the distribution of the admixture which gradually formed as a result of the repeated passage of the zones, can be described for the first passage of the zone (Fig. 28) of the ingot by the equation derived by Pfann:

$$C_1(x) = C_0\left[1-(1-k)\exp\left(\frac{-kx}{b}\right)\right] \quad (31)$$

where C_0 is the initial concentration of the given admixture in the entire volume of the ingot; b is the width of the zone; $C_1(x)$ is the new concentration of the admixture in the solidified part of the ingot in the area x after the single passage of the zone. In the case with $x = 0$, $C_1(0) = kC_0$.

For multiple zone melting, on the basis of the law of the conservation of mass, derived by Reiss and Lord, it was possible to derive an equation for calculating the concentration in the crystal after the n-th passage of the zone. The equation proposed by Reiss and Lord defines the relationship between the change of the concentration of the admixture in the advancing zone and the difference between the flow of the admixture arriving in the zone and the flow of the admixture leaving the zone. The equation was derived using the following simplifying assumptions: the width of the zone b does not change, the cross-section of the zone is constant, k is constant, there is no change of the volume during solidification, mixing of the melt is perfect and there is no diffusion in the solid phase. The final equation has the following form:

$$C_n(x) = \frac{k}{b}\left[\int_0^{x+b} C_{n-1}(x)dx - \int_0^{x} C_n(x)dx\right] \tag{32}$$

but is not valid for the entire length of the ingot. When the melting front of the zone reaches the end of the ingot, this final residue of the melt is characterised by directional solidification.

The comparison of the concentration profiles of zone melting and directional solidification for $k < 1$ indicates (Fig. 29) that the degree of depletion in the admixture in the narrow zone in zone refining is, as a result of the smaller volume of the zone, lower in comparison with the mother liquid can spread widely along the entire length during directional solidification. Consequently, after the single passage of the zone, the level of the concentration of the foreign admixtures in the ingot subjected to a single zone of the melting is higher in comparison with directional solidification. If the width of the zone b were equal to the total length of the ingot l_0, the equation of zone melting (31) changes in the form to the equation of directional solidification (30).

Figure 30 shows the increase of the refining effect after 1–20 passages of the zone. Comparison with directional solidification and its concentration profile indicates that zone melting is an efficient method of refining only if the process is carried out as a multiple refining process. Depending on the parameters of b, k and l_0, it is possible to obtain the final or limiting distribution [7].

A shortcoming of zone melting and also directional solidification when used as the refining method is the joint presence of the impurities with $k > 1$ and $k < 1$ in the main substances. Therefore, prior to selecting these refining methods, it is necessary to evaluate the presence of all impurities in the main substance and their possible distribution in relation to the values of the distribution coefficients or select a different method of removal of these impurities.

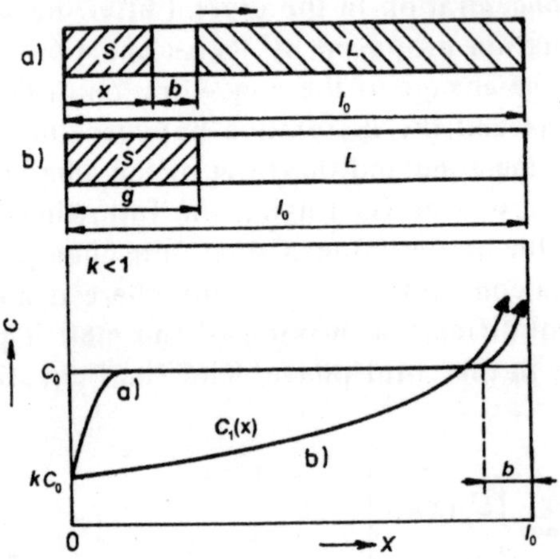

Fig. 29 Comparison of the concentration profiles after a single passage of the zone for $k < 1$. a) in zone melting, b) in directional crystallisation.

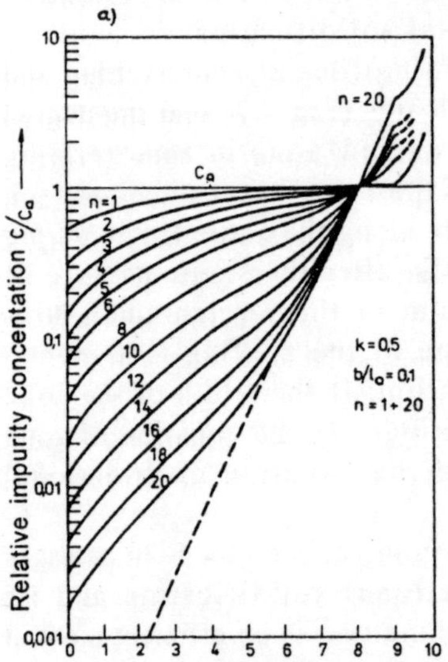

Fig. 30 Concentration profiles for zone melting after 1–20 passages of the zone including final distribution.

After completing the entire *n*-stage process of zone melting, the cleaner, refined part of the ingot is separated by sectioning and sent for further processing.

A greater effect may be obtained by the so-called zone refining with the removal of material in which the final small part of the zone refined substances is gradually removed, for example, after every pass of the zone, from the end of the ingot (at $k < 1$). This zone is characterised by the highest degree of enrichment with admixtures, Fig. 31 [7]. Therefore, this method makes it possible to increase the yield of refined material.

Zone melting is used, in addition to the production of high purity substances, also for the preparation of microalloyed parts of the zone refined ingot of zone grown single crystals or for growing single crystals of high purity substances (silicon, germanium, refractory metals, etc.).

A suitable example is represented by the experimental results obtained for zone refined copper. In the determination of the effective distribution of admixtures in zone refining of copper and for analysis of this process, it is necessary to know the concentration profiles of the distribution of the impurities and admixtures along the length of the zone-melted ingot. The most suitable method for the determination of the concentration profile after zone refining in which very low concentrations of the admixtures are used is the application of selecting radionuclides as a marker admixture. For example, this method was used by the authors of the book to investigate the distribution of radionuclides ^{124m}Sb and ^{110m}Ag in copper [25] and ^{95m}Zr and ^{110m}Ag in aluminium [6,32].

The starting material was copper of commercial purity 4N5, in which the radionuclide of antimony in the amount of approximately $5 \cdot 10^{-3}$ wt.% was added as a marker admixture. Zone melting was carried out in an induction high-frequency equipment in a graphite boat placed in a silica tube in the shielding atmosphere of purified argon. The specimen was subjected to triple zone remelting with a mean travel speed of the zone of 19 cm h^{-1}; after each remelting special radiometric equipment was used to measure the distribution of the radionuclide

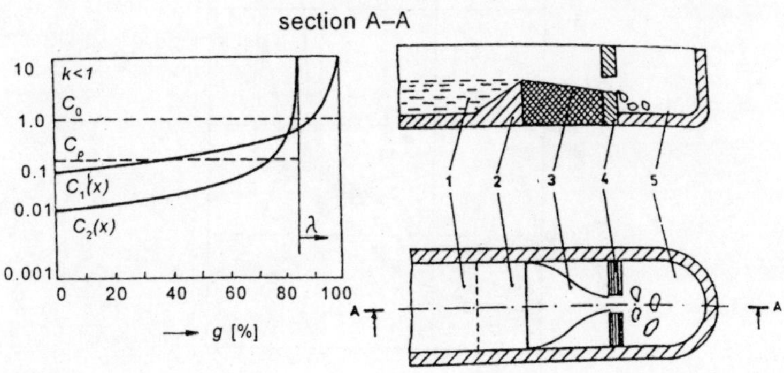

Fig. 31 Principle of directional crystallisation with removal of the material and experimental set-up: 1) melting space, 2) discharge, 3) trough, 4) intermediate wall with capillary orifice, 5) collection space.

of antimony in 1 cm intervals from the upper and lower side of the ingot along its entire length. Figure 32 shows the concentration profile of the ^{124}Sb radionuclide after individual passages of the zone. The concentration curves were process by the Burris–Stockman–Dillon method [38], based on the application of equation (32), to calculate the mean value of the effective distribution coefficients of the radionuclide of antimony in copper $k^{Cu}_{ef\,Sb} = 0.35 \pm 0.06$ which, according to the Burton–Prim–Slichter equation (27), confirms the theoretical assumption $k^{Cu}_{o\,Sb} = 0.17$ (see Table 6).

In zone refining, with respect to the length of the remelted ingot, the admixtures with $k > 1$ always buildup at the beginning of the ingot, Fig. 33, but the admixtures with $k < 1$ buildup always at the end of the ingot, which was confirmed not only in the previously mentioned experiment with the distribution of ^{124m}Sb in the zone refining of copper, but also generally. As the difference between the distribution coefficients and unity increases, the intensity of the distribution process in the macrospace of the ingot increases.

In the presence of a dendritic microstructure, generally in real ingot and castings, admixtures with $k > 1$ accumulate at the axes of the dendrites, and impurities with $k < 1$ accumulate in the inter-dendritic spaces. However, depending on the restricted dimensions of the dendrite arms, their concentration profiles are considerably

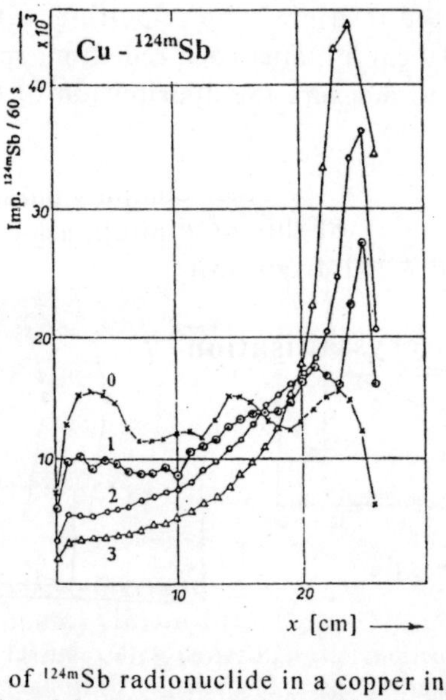

Fig. 32 Distribution of ^{124m}Sb radionuclide in a copper ingot after horizontal zone melting after 0, 1, 2 and 3 passes of the zone (for $k < 1$).

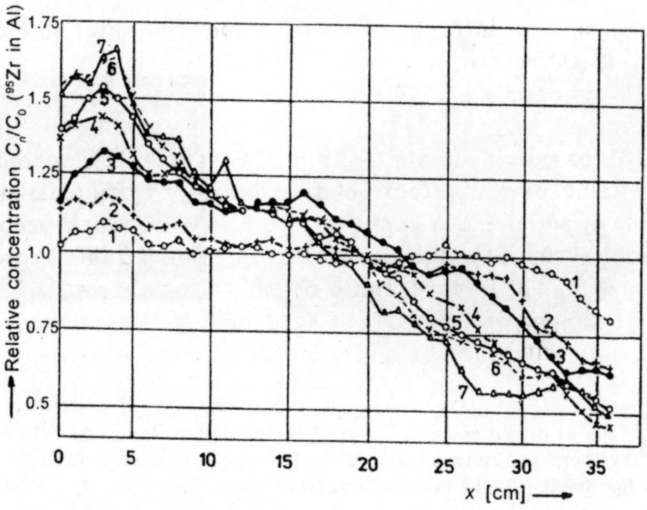

Fig. 33 Distribution of radionuclide ^{95m}Zr in aluminium after horizontal melting after 1 to 7 passages of the zone (for $k > 1$) [6, 39].

less distinctive and are normally identified by metallurgists as the well-known dendritic segregation microinhomogeneities by means of microprobe analysis, autoradiography or other microstructural and metallographic methods – see Fig. 34.

As regards the grains and their dimensions in cast structures, enrichment with admixtures and impurities with $k < 1$ takes place at the grain boundaries and if the degree of enrichment exceeds the concentration of the solid solution α of the given admixture in the main substance, the enriched admixtures effectively precipitate from this substance either as eutectic or other complex phases at the grain boundaries, and the intensity of this precipitation increases with an increase of the value of $k \ll 1$ from unity.

6.7. Mass transfer in crystallisation

In addition to the change of concentration, described by the distribution coefficient k, the melt–crystal interfaces also characterised by a change of the density of the main material of the matrix.

At normal pressure, the volumes of the substances V and their density ρ are a function of temperature. With increasing temperature, the volume V_S of the majority of typical metals (aluminium, copper, iron, etc.) increases with a decrease of density ρ_S. With the change of the solid–liquid state, the volume of the metal V_S and the density

Fig. 34 Dendritic structure. a) Dendrites from an ingot of low alloy steel, b) autoradiograph of distribution (10^{-3} %) of radionuclides ^{59}Fe in Al with purity 4N (k^{Al}_{oFe} = 0.03). Increased concentration of Fe is found in inter-dendritic spaces (white areas) of the dendritic and cellular structure.

of metal ρ_S changes abruptly to V_L and ρ_L, with a further linear change in the melt. In the case of semiconductors germanium, silicon, and elements Sb, Ga, Bi and in some semiconductor compounds (GaAs, InAs, AlSb), as in the case of water, the change of the state $L \rightarrow S$ increases the volume and decreases their density abruptly, Table 7. The physical processes of the change of the volume and density are reversible [40, 41].

Mathematically, the relative change of densities ρ_S and ρ_L and volumes V_S and V_L at the melting point can be expressed by the dimensionless mass transfer coefficient

$$\alpha = \frac{\rho_S}{\rho_L} = \frac{V_L}{V_S} \tag{33}$$

This coefficient α is higher than unity ($\alpha > 1$) for the majority of metals and lower than unity ($\alpha < 1$) for semiconductor substances, Table 7, and characterises the mass transfer which takes place in all crystallisation processes. The same Table gives the relative change of the volume of the substances $\dot{\alpha}$ with the change of the state L–S:

$$\dot{\alpha} = \frac{V_L - V_S}{V_S} \cdot 100 \quad [\%] \tag{34}$$

In horizontal zone melting or directional solidification which is affected uniformly by gravitation, mass transfer takes place and affects the shape of the ingot. A decrease of the volume in melting, i.e. at $\alpha < 1$, results in the displacement of mass in the direction of movement of the zone. On the other hand, the increase of the volume during melting, i.e. in the majority of metals with $\alpha > 1$, results in mass transfer in the opposite direction, i.e. against the direction of movement of the zone. If the molten zone moves in an open horizontal boat, the ingot is usually wedge-shaped. This change after the single passage of the zone is small, but after repeated passages it may become considerable, Fig. 35a [7]. This may also result in overflowing of the material at the end of the boat. However, in certain cases, this process may be desirable, for example, in the zone melting with removal of the material, Chapter 6.6, Fig. 31. However, in the majority of cases it is necessary to prevent mass transfer. This can be carried out by, for example, deflecting the boat with the ingot being melted from the horizontal plane by the critical angle θ_{cr}, Fig. 35b. As a result of deflection, the effect of the parameter α is eliminated and the resultant ingot is characterised by the constant cross-section along the entire length.

The critical angle of deflection of the boat θ_{cr} in the case of zone melting can be expressed by the equation:

$$\theta_{cr} = \text{arctg}\frac{2h_0(1-\alpha)}{b} \qquad (35)$$

and in the case of directional solidification

Table 7 Coefficients α of mass transfer and relative change of the volume of substances $\dot{\alpha}$ with the change of the state $L \rightarrow S$

Substance	$\alpha > 1$	$\dot{\alpha}$ [%]	Substance	$\alpha > 1$	$\dot{\alpha}$ [%]	Substance	$\alpha < 1$	$\dot{\alpha}$ [%]
Li	1.015	1.5	HgSe	1.068	6.77	Si	0.895	−10.5
Na	1.025	2.5	HgTe	1.012	1.23	Ge	0.944	−5.56
K	1.026	2.6	CdTe	1.004	0.4	Ga	0.968	−3.24
Rb	1.025	2.5	CuI	1.09	9.01	Bi	0.967	−3.3
Cu	1.043	4.25	Ga_2Te_3	1.05	5.05	Sb	0.991	−0.95
Ag	1.034	3.4	In_2Te_3	1.019	1.9	AlSb	0.871	−12.9
Au	1.05	5.04	GeTe	1.067	6.7	GaSb	0.904	−9.6
Mg	1.041	4.1	SnTe	1.049	4.9	InSb	0.875	−12.5
Zn	1.047	4.7	PbTe	1.031	3.1	GaAs	0.893	−10.7
Cd	1.047	4.7	PbSe	1.062	6.2	InAs	0.929	−7.1
Hg	1.035	3.5	PbS	1.088	8.8	Mg_2Sn	0.975	−2.5
Fe	1.044	4.4	Bi_2Te_3	1.036	3.58	Mg_2Pb	0.977	−2.3
Al	1.063	6.26	Bi_2Se_3	1.041	4.1			
Tl	1.033	3.26	Sb_2Te_3	1.032	3.18			
Pb	1.034	3.38	Sb_2Se_3	1.092	9.18			
Se	1.18	18	Sb_2S_3	1.131	13.1			

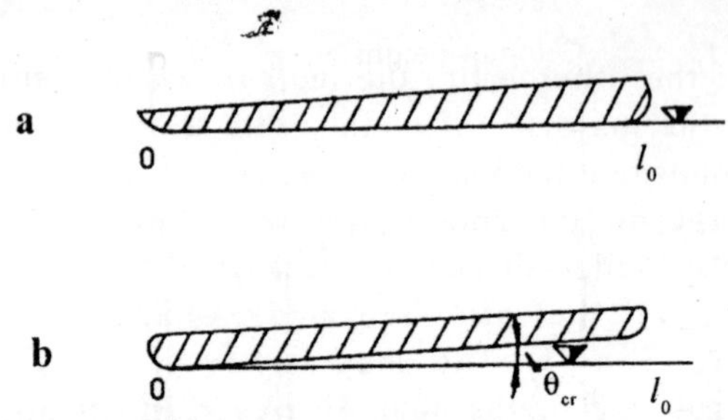

Fig. 35 a) Displacement of mass in directional solidification and zone melting for substances with $\alpha < 1$ (Ge, Si, Bi) after the multistage process with build-up of the substance on the end of the ingot; b) The boat inclined under critical angle θ_{cr} (preventing wedging of the ingot).

$$\theta_{cr} = \operatorname{arctg}\frac{2h_0(1-\alpha)}{l_0} \tag{36}$$

where h_0 is the height of the melt in the boat.

When calculating θ_{cr}, the convection and surface tension of the melt which try to make the surface of the melt round are ignored.

In the case of vertical arrangement in a closed system (melting in a vessel) or melting by the floating zone (FZ) method (section 6.8.3), mass transfer also takes place. This may result in cracking of the closed end of the vessel. For materials those volume decreases after melting ($\alpha<1$), the recommended direction of melting in the vessel is from bottom to top, Fig. 36a. When melting materials with $\alpha > 1$, the direction of melting in the vessel should be from top to bottom, Fig. 36b. In melting by the FZ method, the shape and stability of the zone are affected, in addition to the changes of the density of the melt α, also by the combined effect of gravitation and surface tension.

6.8. Crystallisation methods of refining and production of single crystals of metals, alloys and special materials

6.8.1. General classification

There are more than 100 technologies used for refining different materials, for the preparation of crystals of metals, non-metals, their compounds and alloys with the required content of alloying elements [8,9]. Crystalline and amorphous substances can be prepared from the gas, liquid or solid phase. Crystallisation from the melt is usually preferred due to a very large number of advantages because this method can be used to produce large crystals defined by their external shape,

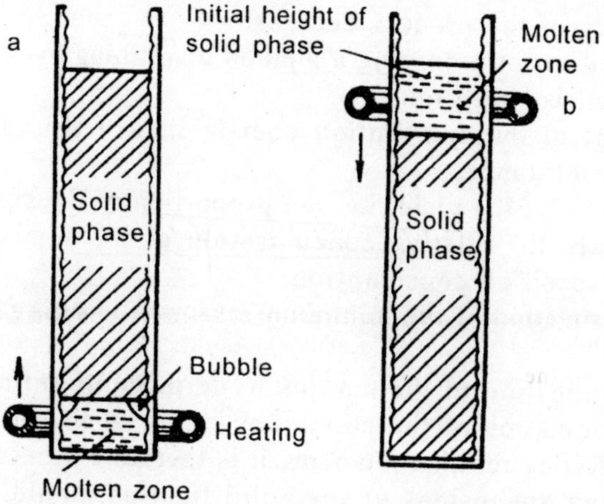

Fig. 36 Recommended directions of movement of the zone in vertical zone melting in the vessel. a) after melting the volume of the melt decreases ($\alpha < 1$), b) after melting the volume of the melt increases ($\alpha > 1$).

dimensions, the degree of chemical purity and structural perfection. In the majority of cases, the crystals are used directly as the initial materials for further technological processing (for example, production of integrated circuits, electrical engineering, electronic or optoelectronic elements). The crystallisation methods, based on the principles of directional solidification and zone melting, are used in production for the refining of the main substances to the maximum possible purity and for the production of crystals with the homogeneous or given distribution of the admixtures.

When designing equipment for refining of substances by the crystallisation methods, it is therefore necessary to propose the optimum parameters which affect the degree of separation of the admixture and the time required for obtaining this degree, selecting a more suitable shape of the solidification vessel, the method of heating and cooling, and the mechanism of the process. For example, in refining by zone melting, to obtain the required yield of the material of the given purity in the shortest possible time and with the minimum expenditure, it is necessary to select the number of passages of the zone n, the length of the zone b, the distance between the zones, the speed of movement of the zone v and the method of stirring the melt. The small length of the zone and the small distance between the zones reduce the time required for individual passages. As a result of an increase of the speed of movement of the zone, the effective distribution coefficient k_{ef} in the macrospace of the ingot approaches 1 and, consequently, the efficiency of the process decreases.

The procedure for proposing a method of refining by crystallisation from the melt is based on:

- knowledge of the distribution coefficients of the admixtures in the main substances;
- the selection of the yield, i.e. the proportion of the material which should have the relative concentration of the admixtures lower than the specified consumption;
- the determination of the minimum ratio n/v, ensuring the required yield;
- the determination of the maximum permissible refining time;
- the determination of the maximum length of the ingot l_0.

When selecting refining methods, it is therefore necessary to select the shape and dimensions of the solidification mould, the heating method, the arrangement of the process, the displacement mechanism, the method of mixing the melt, and the suitable atmosphere or vacuum.

Classification of the methods of refining and production of crystals

a) on the basis of the position of the solidification mould or crystal
   - horizontal (Fig. 37b)
   - vertical

b) according to the size of the volume of the molten charge
   - directional crystallisation – the Bridgman method (Fig. 37a)
     – the Czochralski method (Fig. 37c)
   - the zone melting – in a solidification mould (Fig.37d)
     – the floating zone (FZ) method (Fig. 37e)

c) according to the atmosphere used
   - vacuum
   - inert atmosphere (He, Ne, Ar, N_2)
   - reaction atmosphere (H_2, N_2, O_2, CO, S_2, As_2, Cl_2, I_2)

d) according to the pressure in the melting chamber
   - a very low pressure (a high vacuum)
   - reduced pressure
   - atmospheric pressure
   - high-pressure (melting in an autoclave)

e) according to the material of the solidification mould (in order to prevent contamination of the crystal)
   - the crucibleless method (FZ)
   - the autocrucible (the crucible is made from the same material

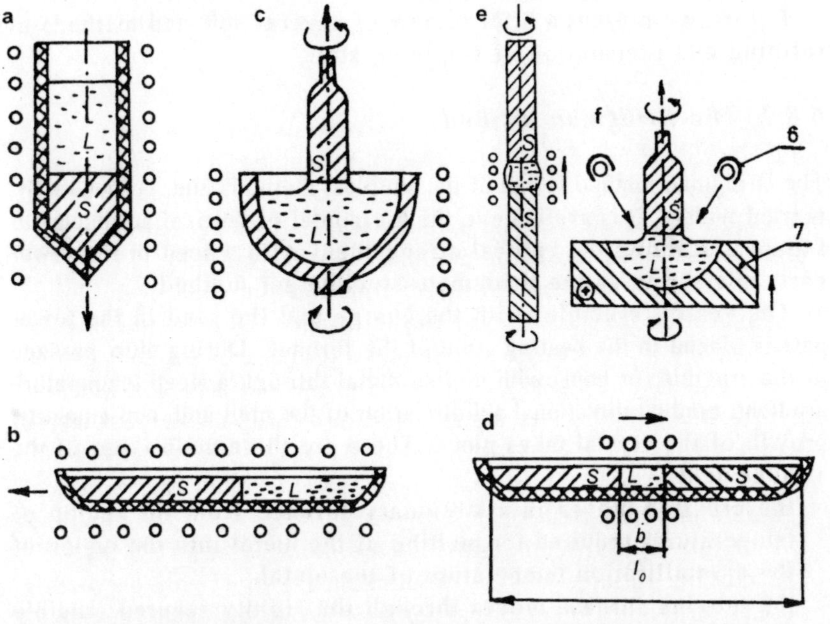

Fig. 37 Methods of refining and production of crystals from the melt.

as the crystal (Fig. 37f)

- oxide materials (quartz, Al_2O_3-corundum, MgO, CaO, ZrO_2...)
- graphite (for materials which do not form carbides)
- copper (copper-plated, water-cooled boat or solidification mould)
- boron nitride (for semiconductor materials)
- noble metals (platinum, iridium)

f) according to the heating method

- resistance heating
- induction (high frequency and low frequency heating)
- gas heating
- electric arc heating
- plasma heating (low-temperature plasma, high-frequency plasma)
- electron beam heating (the beam of accelerated electrons)
- ion heating (ion beam)
- heating by radiation (solar furnaces, laser)
- freezing-out

g) according to the mixing method

- the rotation of the crucible and the crystal
- induced high frequency current
- mechanical
- magnetic, electromagnetic.

Below, we present a brief review of several selected methods of refining and preparation of single crystals.

6.8.2. The Bridgman method

The Bridgman method is one of the simplest methods and, consequently, is used widely. It is available in the horizontal or vertical arrangement. Figure 37a shows the vertical arrangement with a seed in the lower part, referred to as the Bridgman–Stockbarger method.

The vessel (crucible) with the charge and the seed in the lower part is placed in the heating zone of the furnace. During slow passage of the crucible (or boat) with molten metal through a steep temperature gradient, gradual directional solidification of the melt and, consequently, growth of the crystal takes place. There are three main types of the arrangement:

- the crucible moves in a stationary furnace from the region of temperatures required for melting of the metal into the region of the crystallisation temperature of the metal.
- the moving furnace moves through the rigidly secured crucible (boat); this is efficient in particular for operation in vacuum.
- the furnace and the crucible with the melt do not move. Gradual directional crystallisation is achieved by a controlled decrease of the temperature of the furnace (the method of the horizontal gradient freezing or the method of electrodynamic gradient freezing).

6.8.3. Zone melting methods

There are two main arrangements of zone melting: horizontal or vertical.

Equipment for refining of metals by horizontal zone melting in a boat consists of three sections: a boat with a charge, a heating source capable of forming a narrow molten zone in the charge, and a mechanism for moving the boat or the heating source. The material of the boat and the atmosphere of this furnace must not contaminate molten metal. In single crystal production, an oriented seed is placed in the boat and it is necessary to maintain the constant length of the zone during the entire passage or, in some cases, repeated passages of the zone through the specimens. In order to maintain the constant length of the zone, there should be a steep temperature gradient at both ends of the molten zone.

The heating source include resistance heating, high frequency induced current, electric arc, electron beam accelerated by the electric field, the flux of photons (laser), focused solar energy (solar heating), etc.

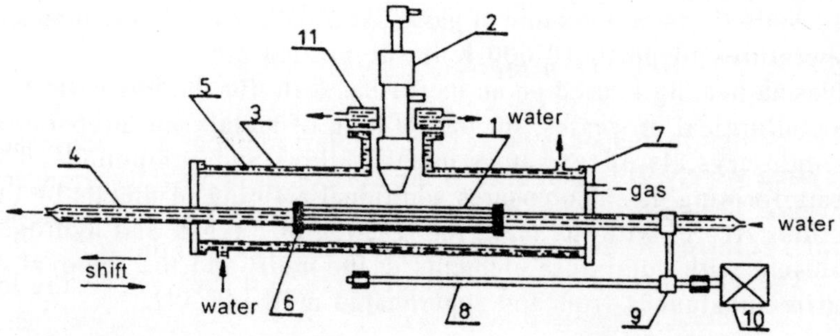

Fig. 38 Equipment for plasma melting of metals. 1) solidification mould, 2) plasma torch, 3) external jacket, 4) carrier pipe, 5) internal jacket, 6) front end of the solidification mould, 7) furnace flange, 8) guide screw, 9) carrier, 10) electric drive, 11) cooler.

The zone melting of refractory metals is carried out in most cases by electron or plasma heating resulting in higher temperatures and enabling efficient control of the process. Induction heating is the most suitable method for metals with medium melting points. Homogeneity of the melt can be ensured by electromagnetic stirring.

Plasma heating is used in the production of special alloys and, in particular, in the metallurgy of high-purity alloys. Plasma forms as a result of thermal ionisation of the gases – argon, nitrogen, hydrogen and other inert gases in the electric arc. Generators of low-temperature plasma include plasma torches working with the electric arc concentrated by a nozzle with the joint aerodynamic effect of the flowing gases. In 1967, the VSB-TU of Ostrava (Czech Republic) developed [42] a plasma torch with the continuous stabilisation of the electric arc running between a water-cooled cathode produced from thoriated tungsten and a copper nozzle used as an anode. Other types of plasma torch (plasmatrons) are based on the ionisation of gas in the high frequency field operating with frequencies of up to several tens of MHz.

The Department of Non-ferrous Metals, Refining and Recycling of the VSB-TU of Ostrava developed, for special applications in the preparation of high purity metals by zone melting, a plasma furnace with a horizontal, copper, water-cooled solidification mould operating with a low-temperature plasma torch (Fig. 38). The furnace can be used for melting metals with medium melting points (Fe, Co, Ni) and with high melting points (Ti to Mo), and for special high-alloyed alloys (Fe–N, Ni–Ti, Al–Fe–Mo, etc.) which cannot be often prepared by other methods because of the high volatility of components (Cr).

This furnace operates with a slight excess pressure of the atmosphere of the ionised gas or a mixture of gases (Ar, Ar–N_2, Ar–H_2) and produces temperatures of up to 10 000 K in the plasma arc.

Plasma heating is used on an increasing scale for the intensification of metallurgical processes, for the refining of metals and preparation of single crystals of refractory metals, alloys and compounds. The plasma-forming gas also ensures additional refining of metals by the reactions $Ar^{+}+e^{-}$ with the melt, the reaction of oxygen and hydrogen in plasma with admixture elements in the melt, and the removal of volatile substances from the superheated melt [43,44].

Vertical zone melting

Crucibleless zone melting, i.e. the floating zone method (FZ), Fig. 39, is one of the most efficient and widely used methods of refining and preparation of crystals of reactive substances and refractory metals. The narrow molten zone is produced by, for example, high frequency, electron beam, laser and solar heating. Electron beam zone melting in high vacuum using suitably oriented nuclei (seeds) can be used for the preparation of single crystals of refractory metals: W, Mo, Ta, Nb, V, Zr, Ti, Re. The zone is maintained in the suspended state mainly by surface tension forces. The stability of the zone depends on gravitation, surface tension and density of the melt, the concentration of the admixturs and the direction of movement of the zone. The stability of the zone is maintained using an external magnetic field with the so-called support frequency. The gradual melting of layers on the original single crystal bar is carried out in individual stages.

In electron beam zone melting, in addition to the effect of zone refining, a significant role is played by secondary processes. They cause the generation of gases and evaporation of admixtures with a high vapour tension in the conditions of working vacuum of 10^{-3}–10^{-6} Pa. The single crystals of metals, produced by this method, are not perfect, especially when the bar diameter is large. During the growth of the crystal and its solidification, samples contains a steep temperature gradient, the crystal–melt interface is curved and, consequently, mechanical stresses form in the crystal. Nevertheless, this method has been used for the growth of single crystals of W, Mo, Ta and Nb with large, slightly disoriented subgrains and with the dislocation density of less than 10^4 cm^{-2} inside the subgrains.

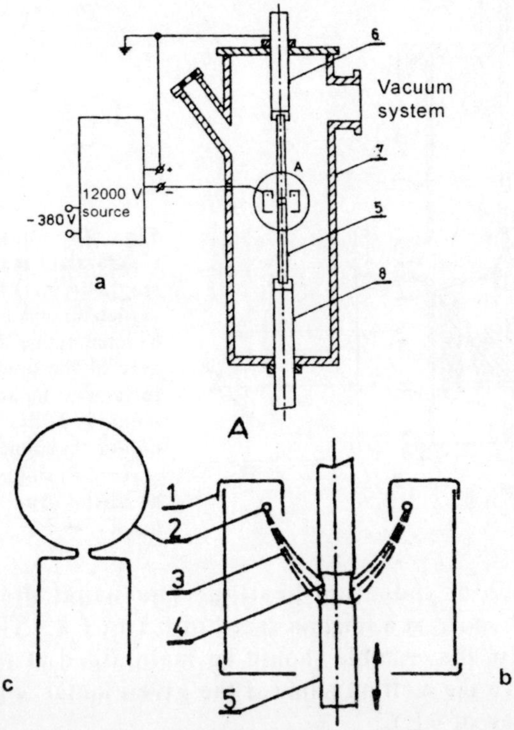

Fig. 39 Equipment for electron beam zone melting in vacuum by the floating zone method. a) section through melting equipment, b) section through focusing system, c) the shape of the tungsten cathode. 1) focusing system consisting of Mo sheets, 2) cathode, 3) electron beam, 4) molten zone, 5) metal used for melting, 6) upper holder of the bar, 7) the jacket of the recipient, 8) lower holder of the bar.

6.8.4. Drawing the crystal from the melt The Czochralski method of single crystal preparation

The method is based on a procedure in which a seed crystal is immersed in the melt in the crucible, the seed is partially melted and, subsequently, is withdrawn from the melt at a lower rate, Fig. 40. The temperature distribution in the cross-section of the crystal must be symmetric in relation to the axis of the crystal; this is supported either by the rotation of the seed or rotation of the crucible with the melt, or rotation of both. The rotation of the crystal and the melt ensures sufficiently high homogeneity of the melt and the minimum width of the diffusion layer at the solidification front. The crucible with the melt is placed inside the closed space of the furnace in vacuum or an inert atmosphere to prevent oxidation of the melt. The heating source is represented

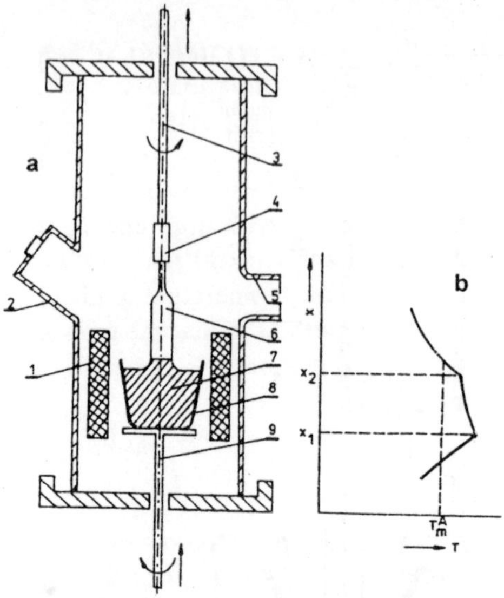

Fig. 40 Equipment for the Czochralski method of drawing single crystals from the melt. a) section through melting equipment, b) temperature distribution in the axis of the crucible, 1) electric resistance heater, 2) inspection window, 3) drawing bar, 4) seed holder, 5) connection of vacuum system, 6) single crystal, 7) melt, 8) silica crucible, 9) crucible holder.

by resistance or induction heating. The usual drawing rate is $1 \cdot 10^{-4}$–$1 \cdot 10^{-3}$ cm s^{-1} at a rotation speed of 0.1 to 1 s^{-1}. The temperature of the melt in the crucible should be maintained at required value slightly above the melting point of the given metal with a relatively high accuracy of ± 1 K.

The temperature distribution along the axis of the crucible in the Czochralski solidification equipment is shown schematically in Fig. 40b. The isotherm, corresponding to the solidification temperature of the substance, is situated above the surface of the melt. The temperature in the crucible increases in the direction to the bottom of the crucible. The single crystal seed and the growing crystal are used for the efficient removal of heat from the solidification front. The form of the surface of the melt is affected by gravitation and the surface tension of the melt. The rate of withdrawal of the crystal, the speed of rotation and the temperature of the melt in the crucible also affect the shape of the crystal–melt phase boundary. At a high temperature of the melt or a high withdrawal speed, the solidification isotherm is situated above the surface of the melt resulting in a decrease of the diameter of the crystal. On the other hand, a decrease of the temperature of the melt and the speed of withdrawal increases the diameter of the crystal which is utilised in practice.

The Czochralski solidification method can be used for producing large single crystals from the melt (in the case of silicon, the diameter to 30 cm, length up to 200 cm) and also the required geometrical shape. The preparation of dislocation-free single crystals of silicon is carried out using the so-called Dash method (formation of a neck) when the annihilation of dislocations from the seed takes place in the narrow part between the seed and the actual crystal. The preparation of dislocation-free crystals or crystals with a low dislocation density should be carried out using the initial material with high chemical purity, a perfect seed, a low growth rate, a small axial temperature gradient, a flat phase boundary, and a minimum temperature gradient in the radial direction. The advantage of the method is that the growth conditions can be controlled quite easily, regulated and also automated. The crystal grows in the free space so that there are no problems with contamination and difficulties associated with the growth of single crystals in the crucible or boat (formation of mechanical stresses in the crystal during solidification and cooling, difficult handling of the crystal, etc). The concentration profile of the admixtures along length of the crystal is identical with the concentration profile in directional solidification. The crystals prepared by the Czochralski method are, however, often characterised by a layered structure, i.e. transverse growth chemical inhomogeneity, resulting from the fluctuation of the temperature of the melt at the phase boundary caused by the problems with temperature regulators or uplift convection.

6.8.5. Preparation of single crystals with a homogeneous or given distribution of admixtures

Single crystals with the homogeneous distribution of admixtures or alloying elements in the axial direction in the macrovolume of the ingot in the directional solidification can be obtained basically by two methods based on:

- maintaining a constant concentration of the admixture in the vicinity of the solidification front under the condition k_{ef} = const.
- programmable change of k_{ef} in the process of single crystal growth.

In the first method, the constant content of an element in the melt can be ensured by external supply from the solid, liquid or gas phase, by means of the composition calculated in advance. The floating crucible method is used in the Czochralski method (Fig. 41) on the basis of the procedure in which the constant concentration of the elements in the internal vessel, in which solidification takes place, is ensured as a result of constant supply of the melt from the external crucible. The concentration of the admixture in the external and internal crucibles and the ratio of their dimensions are determined by the value of k_{ef}. This makes it possible to obtain the constant concentration of the

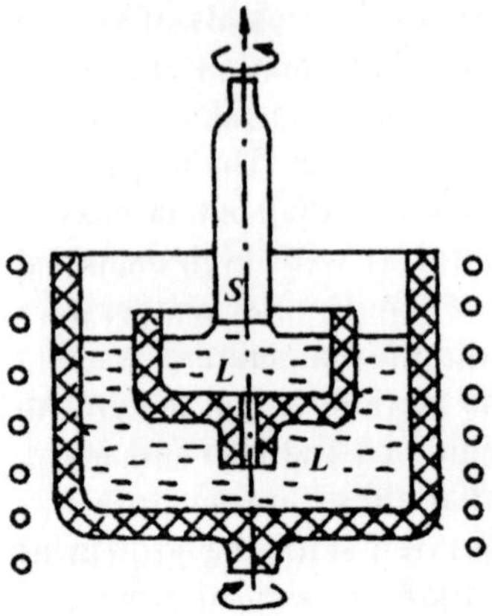

Fig. 41 The floating crucible method.

admixture along the entire length of the crystal. A similar procedure can be used in homogenising in zone melting (the method of the starting charge).

The second method is based on the Burton–Prim–Slichter equation (27), according to which the value of k_{ef} can be affected by the change of the solidification rate and the speed of rotation of the crystal and the crucible, i.e. forced convection in the melt. Selection of the programme for obtaining a homogeneously alloyed crystal in the axial direction is based on the determination of the dependence of the rate of solidification and rotation speed on the drawing time or the length of the crystal. This method can be efficiently combined with the floating crucible method where by selecting the suitable form of the internal crucible it is possible to produce single crystals with a programmed distribution of elements. A similar procedure is used for the preparation of oriented dendritic ribbons of semiconductors produced by rapid withdrawal from the melt.

6.8.6. Verneuil method

This method belongs in the group of the oldest solidification methods for the preparation of single crystals of oxides, in particular ruby, sapphire, etc. It is still used extensively in industry. A charge in the form of a fine powder is placed in a hopper with a vibration feeder situated above the upper part of a plasma torch. The hopper is connected through a funnel with the system of nozzles of an oxygen–hydrogen torch. The powder is regularly discharged through the axis of the nozzle. As a result of heating by the flame, the powder melts and falls on the seed heated by the flame. The rotational and translational movement of the holder of the seed ensures the required growth rate of the single crystal. The surrounding furnace is used for gradual cooling of the grown single crystal.

The selection of materials is limited mainly to oxide materials, in particular of ceramic nature. Despite improvements, the limit of the structural perfection of produced single crystals is lower than, in, for example, the method based on photon heating. This heating is based on the focusing the beam of a high-power light source (plasma torch or sun) to the solidification space.

6.8.7. Preparation of single crystals from the gas phase, the vapour–liquid–solid (VLS) method

A droplet of a solvent is placed on a crystal substrate and at the given temperature, the droplet is additionally saturated by means of a chemical transport reaction (see section 7.8), resulting in the formation of a supersaturated solution. The resultant concentration gradient leads to diffusion in the droplet and to crystallisation on the substrate. The result is the growth of a sharp-tip 'whisker single crystal', with the tip of the crystal covered by the solvent.

The first stage of growth of this crystal is the formation of nuclei. In the preparation of crystals from the gas phase, the condition for the growth of nuclei is the essential supersaturation of the vapours. According to the current views regarding crystal growth mechanisms, the bonding of atoms to the crystal takes place on steps formed by a screw dislocation. In this case, the formation of a closely spaced plane is never completed and results in spiral growth of the crystal, i.e. a whisker crystal. Theoretically, a single screw dislocation is sufficient for the growth of a crystal at low supersaturation. Crystallisation from the gas phase was used to obtain whisker crystals of Ge, Si, Fe, Sn, Zn, Ag, Cu, Ni, etc, with the mechanical strength close to the theoretical value.

The single crystals of refractory metals with a low vapour tension are efficiently produced from the gas phase of their volatile compounds. This is carried out by chemical transport reactions of compounds in the gas phase (Chapter 7) on the surface of the crystal substrate. This method has also been used to produce, for example, single crystals of W, Mo, Nb, Ta and other refractory metals. The initial material are halides of these metals, for example, chlorides or fluorides. In this method, the chloride or fluoride of the appropriate refractory metal is placed in the evacuated space and heated to transfer it to the gas phase. Using a hydrogen flow, the vapours of tungsten hexachloride WCl_6 are supplied to the heated tungsten fibre (approximately 1000 °C) where WCl_6 is reduced. The reduced tungsten settles on the heated fibre. When selecting suitable cooling rates, temperature and concentration conditions, a single crystal grows on the seed. As a result of the combined reduction of the mixtures of chlorides, it was possible to obtain alloys of the type W–Mo, Mo-Nb, W–Nb and superconducting layers of Nb_3Sn on a metallic substrate.

6.8.8. Preparation of single crystals from the solid phase-recrystallisation methods

The physical principle of crystallisation is based on overcoming atomic bonds in the deformed crystal lattice, the formation of nuclei of new grains and subsequent growth by the transfer of atoms from the disrupted areas to thermodynamically more stable positions as a result of the formation of relatively perfect crystals. Recrystallisation is accompanied by a decrease of internal stresses. The rate of grain growth depends directly on the initial purity of metal. The grain size after recrystallisation annealing depends greatly on the degree of prior plastic deformation. The maximum growth of the grains is recorded at low, so-called critical degrees of deformation (2–10%), and also at deformations higher than 90%.

The recrystallisation methods of preparation of single crystals in the form of fibres, wires and ribbons can be divided into:

- the methods utilising the polymorphous transformation of the metal in the solid state Fe (see Appendix), U, Ti, Zr, lanthanides. In this case, the accompanying phenomenon from the point of different solubility of the admixture in high-temperature C_v and low-temperature C_n phases is the so-called transformation distribution coefficient at $k_{tr} = C_v/C_n$ of the admixtures present in the main substance [6];
- the methods of recrystallisation annealing of metals after the critical degree (5–10%) of plastic deformation (aluminium, tungsten) or after 80–95% deformation (molybdenum, titanium...);
- the methods of secondary recrystallisation utilising the rapid growth of crystal grains at a specific temperature gradient after alloying with suitable high-melting finely dispersed particles (for example, ThO_2 in W...), thus preventing the growth of further parasitic grains.

In comparison with the method of preparation of single crystals from the gas or liquid phase, recrystallisation methods are characterised by advantages and disadvantages. The disadvantages include the fact that it is not possible to obtain crystals of large cross-sections nor single crystals of the required crystallographic orientation and geometrical shape. However, the recrystallisation methods can be used quite easily for the formation of single crystals of alloys of solid solutions in a relatively wide concentration range, even in the case of alloys with a wide range of solidification temperature.

6.8.9. Preparation of crystals in zero gravity conditions

The aim of all solidification experiments in the zero gravity conditions in space is to obtain further information on the effect on hydrodynamics at the crystal–melt phase boundary. This can be obtained by understanding the differences in the mechanisms of transfer phenomenon in the absence of gravity. The differences are reflected in the behaviour of the melt, sedimentation (different density of the melt $\Delta\rho(T, C)$) and different types of gravity-determined convection (uplift convection).

The forces that are independent of gravity include, for example, the change of the volume during solidification $V \approx 1/\Delta\rho_{L-S}$, Marangoni convection resulting from the changes of the surface tension of the melt $\Delta\sigma(T, C)$ and diffusion. All these factors contribute to understanding differences in the mechanism of heat and mass transfer in the formation of different morphologies of the solidification boundary in the zero gravity conditions [14].

6.8.10. Rapid solidification

During cooling from a melt, metals solidify at temperature T^A_m. However, as soon as the cooling rate exceeds 10^{10} K s^{-1}, the mobility of the atoms is theoretically restricted in such a manner that they solidify by the amorphous mechanism, i.e. retain the atomic structure of the melt. There is the solidification temperature T_G ($T_G < T^A_m$) characteristic of this limited mobility of the atoms.

At present, it is possible to obtain cooling rates of 10^6–10^8 K s^{-1} and, consequently, the amorphous state can be produced only in certain metallic systems, for example:

metal–metalloid	T^A_m (K)	T_G (K)
$Fe_{80}P_{13}C_7$	1280	710
$Au_{81}Si_{19}$	620	300
metal–metal		
$Cu_{60}Zr_{40}$	1550	900
$Ni_{60}Nb_{40}$	1580	1020

In rapid solidification, a melt jet is sprayed between two rapidly rotating cooled copper rolls. In this case, the amorphous (microcrystalline) solid metallic ribbon with a thickness of several millimetres and several centimetres wide solidifies between the rolls during several microseconds. In fact, the ribbon is injected into the space behind the rolls. In this case, coiling is the technical problem.

Amorphous metals, the so-called metallic glasses, formed during rapid solidification, are homogeneous materials ($k_{ef} \rightarrow 1$) without grain boundaries and are characterised by high strength and elasticity. In contrast to commercial glasses, they can be compressed by up to 50%. They are characterised by high corrosion resistance, electrical connectivity and can be easily magnetised and, taking into account low electric losses, they already replace conventional transformer sheets. In heating to temperatures higher than recrystallisation temperature, amorphous materials may change to the crystalline state and, consequently, it is important to consider the stability of metallic glasses.

The results of these rapid solidification technologies (in addition to metallic glasses) are also structural variants of microcrystalline materials with the grain size smaller than 1 μm which, like the amorphous materials, are characterised by the following characteristics:

- wider range of the solid solution;
- the limiting solubility of alloying elements and admixtures is displaced without causing precipitation of specific phases;
- refinement of the precipitated phases;
- the formation of metastable phases because of suppression of diffusion processes.

The processes of rapid solidification are used at present for processing not only magnetically soft materials but also for modification of, for example, the properties of aluminium alloys based on superduralumin Al–4.5Cu with 1–3% Li. In the case of aluminium it is also possible to obtain a higher solubility of iron, nickel and manganese (3–8%), the solubility of chromium, zirconium and cobalt in copper is increased, and in steels it is possible to increase the solubility of carbon, boron, silicon and phosphorus. Similarly, for special applications it is possible to produce improved types of dispersion-hardened materials with a higher concentration of ultrafine precipitates. The processes of rapid solidification also indicate the trends for materials with the application of trace elements and rare earth metals (lanthanides) in promising areas of electrical engineering and microelectronic materials.

7. EVAPORATION, CONDENSATION, TRANSPORT REACTIONS

7.1. The theory of separation of compounds by evaporation and condensation

The phenomena of evaporation, sublimation, distillation and condensation are of great importance in the separation and refining of metals and their compounds. They are based on different tension (pressure) of the vapours of the separated elements, the main metal, the compounds and the admixture.

The processes of evaporation and condensation are the main methods in, for example, refining of magnesium, calcium, zinc and other metals characterised by high vapour tension. The industrial methods of separation of liquid chlorides, for example $TiCl_4$ and $SiCl_4$, are also based on their different ability to transfer to the gas state. The processes of distillation and, in particular, the processes of redistillation or rectification, are used in the refining of intermediate products in the production of semiconductor materials, for example germanium (refining of $GeCl_4$). Deep purification to remove admixtures is also carried out by the distillation or sublimation of the main substance in the form of volatile compounds, for example, lower halogenides, which dissociate after the separation of the pure component in different types of transport reactions.

In subsequent sections, the term distillation will refer to the process of evaporation from liquids or melt and sublimation will be the direct transfer of substances from the solid to gas state.

The mathematical expression of the dependence of the pressure of saturated the vapours p of a substance on temperature T results from the laws of chemical thermodynamics and, in the final relationship, is described by the Clausius–Clayperon equation:

$$\ln p = -\frac{\Delta H_{ev(0)}}{RT} + \frac{\alpha}{R}\ln T + \frac{\beta T}{R} + ...I \qquad (37)$$

where $\Delta H_{ev(0)}$ is the molar enthalpy of evaporation of the substance at temperature T = 0 K; α, β are temperature coefficients; I is the integration constant.

For the majority of cases in technology, it is possible to ignore the change of the enthalpy of evaporation with temperature. If $\Delta H_{ev} = \Delta H_{ev(0)}$ (where ΔH_{ev} is the change of the enthalpy of evaporation of the substance at the boiling point), the temperature coefficients α, β can be ignored and equation (37) changes to the following form:

$$\log p = \frac{-\Delta H_{ev}}{2.303\ RT} + I \qquad (38)$$

or, in general

$$\log p = \frac{-\beta}{T} + A \qquad (39)$$

In this equation $B = \Delta H_{ev}/19.15$. The empirical values of B and A can be found in thermodynamic tables.

The activity of the component in a solution is a function of the concentration of this component which, in the overall form, characterises the degree of cohesion of the molecules or atoms of the component. On the basis of the activity it is possible to evaluate the deviation of the properties of the given component of the solution from its properties in the ideal solution at the same molar concentration.

Therefore, the activity a_i of the component i is the value proportional to concentration, i.e. the molar fraction x_i

$$a_i = \gamma_i\ x_i \qquad (40)$$

In equation (40), the term γ_i is the activity coefficient of component i.

According to the Raoult law, this term characterises the deviation of the properties of the examined component in the given solution from its properties in the ideal solution. Since the deviation of the properties of the real solution from the properties of the ideal solution is associated with the activity of the component in the solution, and the activity of the given component in the solution is determined by the vapour tension of this component in the saturated vapours above the solution: as the activity increases, the vapour pressure also increases, and vice versa. Consequently, it is possible to determine the activity of the component as the ratio of the tension of its vapours above the solution p_i to the tension of the saturated vapours of the same component in the pure state p_i^0:

$$a_i = \frac{p_i}{p_i^0} \qquad (41)$$

For the ideal systems, governed by the Raoult law, the activity coefficient is equal to 1. If the activity coefficient is higher or lower than 1, the activity of the component in the solution and the tension of its vapours above the solution correspond to the higher or lower values than indicated by the Raoult law. Using this method, the deviation of the properties of the examined component in the given solution (the vapour tension in the present case) from the properties in the corresponding ideal solution can be expressed either by the activity or activity coefficient. On the other hand, after measuring the vapour tension, it is possible to calculate the activity of the given component in the solution.

In alloys close to the ideal solution, when the curves of activity of their component approximately corresponds to the Raoult law, it is possible to expected the approximate equality of the attractive forces between the atoms of different types.

Figure 42 shows the curves of the vapour tension of metals in relation to temperature, calculated on the basis of equation (39), which may be used to divide the metals approximately on the basis of their vapour tension into four groups:

I. Volatile – mercury, rubidium, cesium, potassium, cadmium, sodium, zinc, magnesium

II. Medium volatility – strontium, lithium, calcium, barium, antimony, lead

III. Low volatility – tin, manganese, chromium, silver, beryllium, aluminium, copper, gold, iron, nickel.

IV. Very low volatility – platinum, molybdenum, niobium, tantalum, tungsten.

From this approximate classification of the metals based on the tension of their saturated vapours it is possible to evaluate the potential capacity for separation of these metals by distillation. The easiest mutual separation takes place in the case of the metals of the group I and IV, whereas the metals of the groups II and III are more difficult to separate. The experiments show that, for example, the metals such as beryllium, aluminium and magnesium, present in the same group, are very difficult to separate during distillation. This also relates to the mutual separation of zinc and cadmium or magnesium from the admixtures of alkaline metals present in them. Refining of these metals by subsequent rectification distillation has been technically mastered.

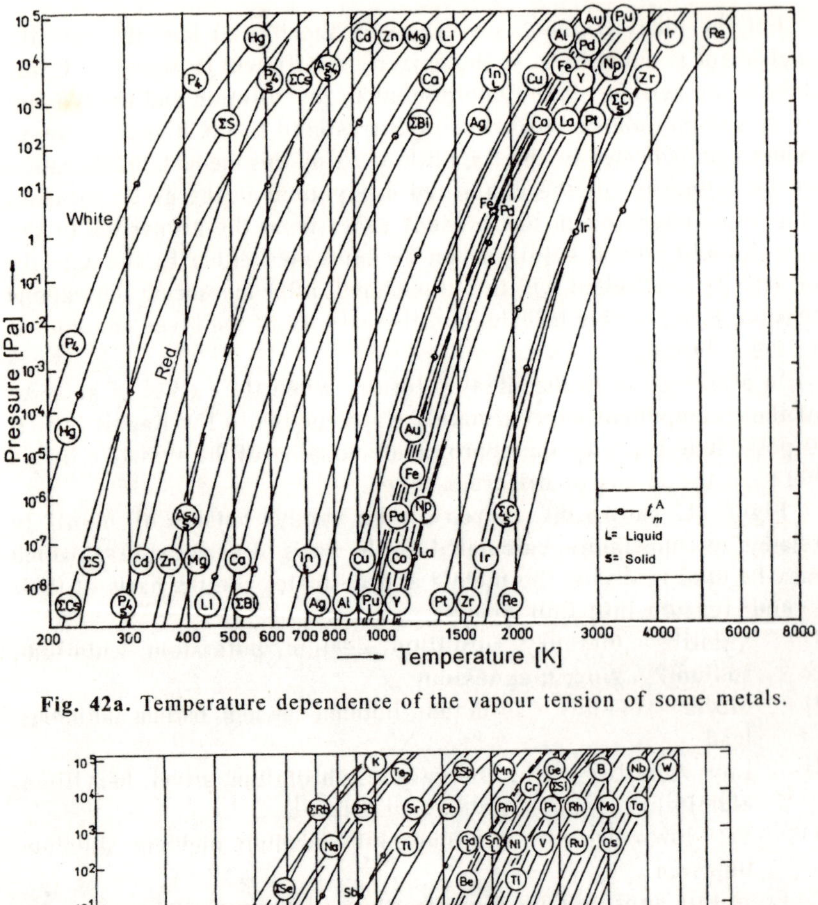

Fig. 42a. Temperature dependence of the vapour tension of some metals.

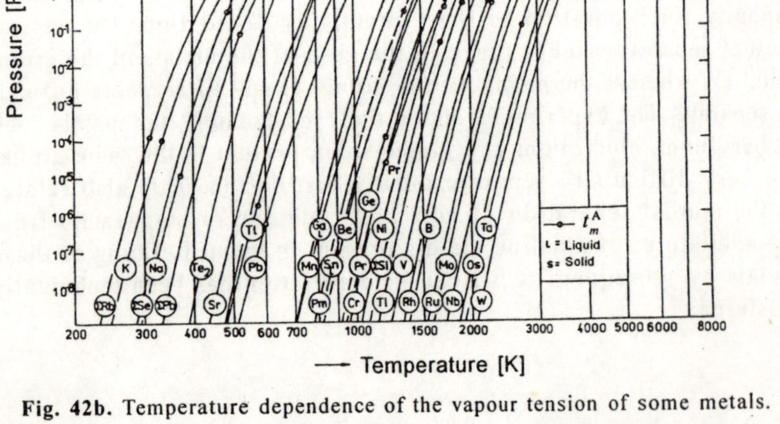

Fig. 42b. Temperature dependence of the vapour tension of some metals.

7.2. The relationship between the composition of the vapour and the solution

The activities of metals in alloys, or the activities of components in solutions, determine the intensity of the condensation processes. It is also important to examine the problem of the relationship between the composition of the vapour above the solution and the composition of the solution, if the latter contains both components. Generally, the relative content of the components in the vapour differs from their average composition in the solution.

Returning to the ideal systems, the composition of the vapour above the solution is equal to the composition of the solution only in the A–B system in which the components have the same vapour tension in the pure condition, i.e. when $p^0_A = p^0_B$. However, this case is exceptional. Usually, the composition of the vapours differs from the composition of the solutions and the difference increases with increasing difference of the tension of the vapours above the pure components. In the simplest systems, the dominant component in the gas phase in comparison with a liquid phase is always the component characterised by the higher vapour tension in the pure condition.

It is assumed that it is component B. Figure 43 shows (dashed lines) the symmetric curves expressing the dependence of the composition of the vapour on the composition of the solution in the ideal system. The diagonal of the square corresponds to the case in which the composition of the vapour and the solution is identical. The deviation from the diagonal increases with an increase in the difference of the pressures of the vapours of the components in the pure condition.

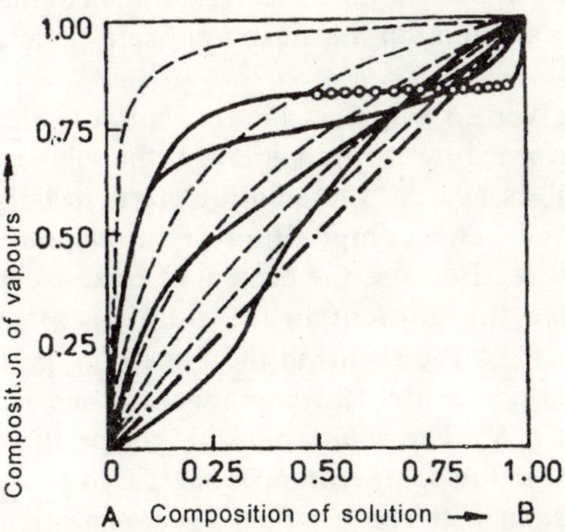

Fig. 43. Composition of equilibrium solutions and vapours in various systems.

In systems characterised by a deviation from the Raoult law, which are characterised by the nonlinear dependence of the composition of the vapour on the composition of the solution, the dependence of the composition of the vapour on the composition of the solution is indicated by complicated curves. These curves are shown by solid curves in Fig 43. The positive deviation of the vapour pressure causes curvature in the upper part of the curves, negative curvature in the lower part. The curvature of the curves increases with an increase of the magnitude of the deviation of the vapour pressure from the linear dependence. The solutions, corresponding to the maximum or minimum points, are referred to as azeotròpic (constantly boiling). They play a significant role in the distillation processes. These solutions are indicated by the points of intersection of the curves with the diagonal of the square in Fig 43.

7.3. Change of the composition of solutions and vapours in relation to temperature and composition–boiling temperature and composition–vapour pressure diagrams

Figure 44 shows the typical diagrams of the composition of the solution–boiling temperature and the composition of the solution–vapour pressure systems. The diagrams, representing the relationship between the boiling point and the composition of the mixture, always contain two curves, one of which corresponds to the temperatures and composition of the liquid phase (melt) and the other one to the composition of the vapour. In this case, the lower curves correspond to the composition of the solution and the upper curves the composition of the vapour. In the composition–vapour pressure diagrams, the position of the curves is reversed. The systems with the minimum on the curves of the boiling points are characterised by the maximum on the vapour pressure curves, and vice versa.

Initially, attention will be given to a system that is similar to the first type (Fig. 45a), i.e. it is governed by the Raoult law. If the solution (melt) with the composition N_1 is heated, the solution starts to boil when temperature t_1 is reached. The composition of the vapour, equilibrium with the solution, is N_3. Because the content of component B in the vapour is higher than the content in the solution, after evaporation of a certain amount of the solution the remaining part of the solution will contain a high content of component A and its composition will be, for example, N_2. The solution of this composition can be brought to boiling only if the temperature is increased to t_2. The vapour which is in equilibrium with the solution, has composition

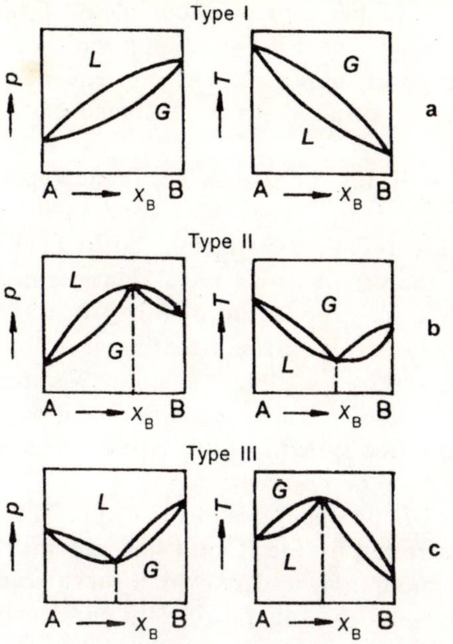

Fig. 44. Pressure–composition and boiling point–composition diagrams for various systems.

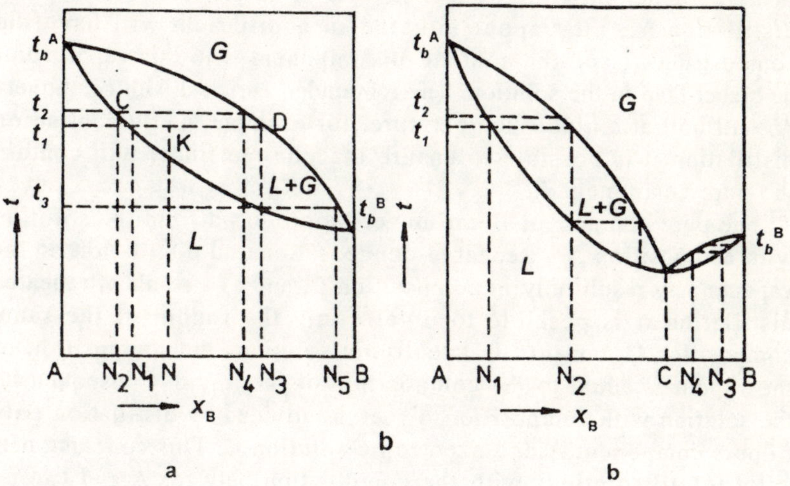

Fig. 45. Change of the composition of the solution and vapours for systems: a) governed by the Raoult law; b) with the azeotropic point.

N_4. It also contains a higher content of component B than the solution. Consequently, the remainder of the solution is enriched with component A and the boiling point increases. The consequence of evaporation is finally the formation of an almost pure component A and the boiling temperature is t_b^A.

The gas phase will now be discussed. If the vapour formed from the solution with the composition N_3 is condensed and the resultant condensate is subjected to distillation, boiling will take place at temperature t_3 and the vapour will have a higher content of component B (composition N_5). The repetition of this process of condensation and distillation may result in the situation in which the separated vapours is represented by the almost completely pure component and the boiling point will the t_b^B. This method makes it possible to separate the binary mixtures in the systems of this type by the method of repeated distillation up to pure components.

In the systems of the second and third type (Fig. 44b and 44c) which are not controlled by the Raoult law and are characterised by a minimum or a maximum, and in which these points are referred to as azeotropic points, the separation into pure components by this method is not possible.

We examine the process of distillation in a system belonging to types with azeotropic point C, i.e. the system has a minimum (Fig. 45b). If the solution with the composition between points A and C (i.e. the solution differing from azeotropic) is subjected to, for example, distillation N_1, the vapour situated in equilibrium will have the composition N_2 and the content of component B in this vapour will be higher than in the solution. The remainder, enriched with component A, will boil at a higher temperature, for example, t_2. By continuing distillation it is possible to ensure that the remainder will contain the pure component A.

The pure component B can never be produced from the solution with composition N_1. Repeated condensation and distillation of the vapours may result only in composition C, and as a result of repeated distillation it is possible to obtain only the vapour of the same composition C, because in azeotropic solutions the composition of the vapour is equal to the composition of the liquid. Consequently, the solution with composition N_1 can be divided by distillation only to pure component A and azeotropic solution C. This conclusion is valid for all solutions with the composition between A and C.

In condensation of alloys whose components in the pure state are characterised by greatly differing vapour pressures, the formation of azeotropic mixtures is less likely and the vapour usually contains an insignificant amount of the volatile substance. In these alloys it is possible to obtain the most efficient separation. On the other hand, the metals with similar vapour pressures in a wide temperature range are more difficult to separate. The alloys of these metals are characterised by higher susceptibility for the formation of azeotropic systems. Almost complete separation of the components is the characteristic feature of the melt of the binary alloys of metals, Al–Zn, Al–Mg, Cu–Zn, Cu–Cd, Cu–Bi, Cu–Li, Sn–Bi, Cu–Pb, Pb–Zn. On the other hand, separation of the melts of the following binary alloys is not possible or is incomplete: Al–Be, Al–Si, Cu–Ni, Cu–Si, Cu–Fe, Fe–Ni, Fe–Co, Ni–Co, Pb–Sn. The alloys Pb–Zn, Cu–Pb and Cu–Fe are characterised by limited solubility in the liquid state and are not governed by the previously described relationships. It is very difficult to separate by distillation the alloys forming, for example, silicides and alloys with intermetallic compounds, for example, Fe–Si, Pb–Ca.

In distillation, the metal vapours are removed from the evaporation space into the condensation space. Condensation starts at a temperature at which the partial pressure of metal in the gas is equal to the partial pressure of the vapour of the pure metal:

$$p_{\mathrm{M}} = p_{\mathrm{M}}^{0} \tag{42}$$

Condensation starts with the formation of condensation nuclei and continues by the growth of these nuclei or by the formation of new ones. The condensation nuclei form by the mechanism in which a temperature decrease is accompanied by a decrease of the kinetic energy of metallic molecules and the contacting molecules remain in the form of liquid droplets acting as condensation nuclei. This phenomenon takes place during a change of energy because the formation of a liquid particle is accompanied by volume work and consumption of energy for the formation of a new surface.

In condensation, it is necessary to ensure that the metal which was originally in the gas phase is transferred back to the liquid (or already solid) condensate with maximum efficiency. The resultant liquid droplets may condense more efficiently if they are larger. The outgoing gas takes away the amount of vapours which decreases with a decrease of the temperature of the gas. If it is required to produce the metal in the liquid phase, the gas should be cooled down to the melting point of metal. However, at this temperature, there may still be a large amount of the metal in the form of vapour and, consequently, the loss of metal increases with an increase in the amount of the outgoing gas from the condenser. If it is required to condense metallic vapours from highly diluted gases, the dew point is very low and may approached the melting point. Consequently, the cooling range is also very narrow and a large part of metal remains in the gas state.

On the basis of the above results, it is possible to summarise the following conditions of efficient condensation:

- the partial pressure of metal vapours should be as high as possible, the dew point should be approximately equal to the melting point;
- the cooling rate of the gas should be low to ensure the formation of larger condensed articles;
- the gas should contain the minimum amount of foreign condensation nuclei.

These principles must be taken into account when designing distillation and condensation systems and in the application of technology of production and refining of metals by distillation methods.

7.4. The physical–chemical nature of sublimation and distillation processes

Sublimation can take place in cases in which the substance heated to a temperature lower than the melting point is characterised by a sufficiently high vapour pressure. In sublimation, the admixtures characterised by higher volatility than the main substance (usually metal), may already separate at a lower temperature, whereas the admixtures with low volatility remain in the remainder after sublimation.

Distillation is a process of evaporation of liquids and makes it possible to separate solutions or melts into their main parts characterised by different vapour pressures. The principle of this refining and separation method is the different composition of the equilibrium liquid and vapours. The quantification characteristics of refining or separation are represented here by the separation coefficients of distillation K_F:

$$K_{FA} = \frac{x_{GA}}{x_{LA}}; \quad K_{FB} = \frac{x_{GB}}{x_{LB}} \tag{43}$$

where x_{LB} and x_{GB} is the molar amount of the substance B in the solution (L) and in vapours (G). As the difference between the values of K_{FA} and K_{FB} and 1 increases, the efficiency of refining also increases.

Because refining of the substances by distillation is accompanied by the redistribution of the main metal A and the admixture B between the gas and liquid phases, the so-called distribution number is of considerable importance for deep refining of substances:

$$b_R = \frac{K_{FA}}{K_{FB}} \tag{44}$$

In distillation of the solutions whose vapours are governed by the laws of ideal gases, i.e. also by the Raoult law, the distribution number b_R may be represented as the coefficient of relative volatility

$$b_{R\,id} = \frac{p_A^0}{p_B^0} \tag{45}$$

where p_A^0 and p_B^0 are the pressures of the saturated vapours of the main substance and the admixture.

In real solutions, taking into account the interaction of the main metal and the admixture, their activity also decreases and, consequently, the pressure of the saturated vapours drops. Consequently, in distillation of real solutions, the distribution number or the coefficient of relative volatility must be expressed by the equation:

$$b_{R\,id} = \frac{p_A^0 \gamma_A}{p_B^0 \gamma_B} \tag{46}$$

Because pure substances are mostly melts in which one atom of admixture B belongs to 10^6–10^9 atoms of the main component A, then without a large error the concentration of the admixture tends to zero and the activity coefficient γ_A may be equal to one. Consequently:

$$b_{R(x \to 0)} = \frac{p_A^0}{p_B^0 \gamma_B} = \frac{b_{R\,id}}{\gamma_B} \tag{47}$$

This shows that the coefficient of relative volatility (distribution number) in distillation of admixture B with a very low concentration is to a certain degree affected by the value of the activity coefficient γ_B of this admixture. However, research shows that the behaviour of admixtures during efficient refining of substances by means of distillation is rarely governed by the laws of ideal gases.

In distillation, the concentration of admixtures in the remaining liquid phase is uniform, especially as a result of the efficient diffusion of admixtures in this phase, and the admixtures therefore do not buildup in the surface layer because it is possible to obtain a very pure main component (with purity higher than in sublimation), without the risk of contamination of the distillate with admixtures. The characteristic feature of distillation is also that it may be carried out as a continuous process whereas sublimation is only a discontinuous, periodic process.

The rate of the sublimation process and of distillation is affected by a large number of factors. The most important factors are the composition of the alloy (content of the admixtures), the size of the surface, temperature and residual pressure. The rate of evaporation increases with an increase of the concentration (activity) of the component in the alloy and with increasing vapour pressure of the component. On the other hand, taking into account the fact that evaporation is independent of the nature of deviation from the Raoult law, providing the activity of the components of the alloy decreases, the rate of evaporation decreases regularly.

Since evaporation takes place from the surface layers, the surface of the metal should be as ragged as possible to ensuring a high rate of the evaporation process. Increasing temperature increases the rate of evaporation of metal but for different components (admixtures) of the melt, the extent of this acceleration may differ. Therefore, with increasing temperature it is necessary to consider the risk of possible contamination of the condensate (sublimate, distillate) with admixtures with lower volatility.

Special attention will be given to the effect of residual pressure. The rate of evaporation, like the rate of any physical chemical process, depends strongly on the rate of removal of the resultant products from the reaction zone. In our case, the residual gases inhibit evaporation because they reduce the rate of diffusion of the gases in the direction of the condenser (the condenser is part of the distillation system in which condensation takes place, i.e. a change of the state of the gas phase to the liquid phase). As the residual pressure and, consequently, the density of residual gases increase, the rate of diffusion of vapours to the condenser decreases and, consequently, the rate of evaporation drops. On the other hand, as the residual pressure decreases, the speed of approach of the evaporating metal to the condenser increases and, consequently, the evaporation rate increases. Therefore, distillation and sublimation should be carried out in vacuum because in this case suitable rates of evaporation for practice are already obtained at relatively low temperatures.

Vacuum has a beneficial effect on the continues removal of metal vapours from the reaction space. The processes of sublimation and distillation are characterised by sufficiently higher rates for practical purposes only when the equilibrium pressure of the vapours of distilled metal is the same or higher than the residual pressure in the system. The main task of vacuum is not only the protection of evaporated metal against oxidation but also to reduce the partial pressure of gases and, consequently, the content of these gases in the metals subjected to vacuum treatment.

7.5. The removal of admixtures by evaporation in vacuum

When examining of sublimation and distillation it is usually assumed that the admixtures are characterised by low vapour pressures or higher boiling point than the main metal. However, cases are often encountered in which distillation or sublimation may be used for the removal of certain admixtures without affecting the main metal. In these cases, the admixtures can be removed from the main metal by heating the metal in vacuum, for example, highly volatile admixtures such as zinc and cadmium from lead, zinc and cadmium from silicon, arsenic and antimony from germanium, etc.

The vapour pressure of the admixtures greatly decreases and the rate of removal of admixtures is lower if they react chemically with the main metal. For example, the vapour pressure of arsenic above its solution in a silicon melt is higher than the vapour pressure of phosphorus above its solution in the silicon melt. This may be explained that the fact that phosphorus forms with silicon a considerably stronger chemical bond than silicon with arsenic, thus reducing the activity of phosphorus in the alloy.

The rate of evaporation of admixtures from the main metal depends on the residual pressure above the solution and the rate of diffusion of the admixtures to the surface of the metal. For efficient removal of admixtures it is necessary to ensure that the residual pressure in the system is (if possible) lower than the equilibrium pressure and the evaporation area should be large.

In certain cases, good results with the removal of admixtures by evaporation are obtained by heating the purified material in a gas stream with which the admixture may react with the formation of a volatile compound and evaporate. Usually, hydrogen is used for these purposes: it reduces many oxides, sulphides, selenides, halogenides. The reduced admixtures evaporate and halogenides, sulphur and selenium are removed in the form of HCl, H_2S, etc. Moist hydrogen is used, for example, for the removal of boron from silicon, whereas hydrogen react with boron with the formation of volatile hydride BH_3. Similarly, by passing the vapours of $SiCl_4$ above molten silicon, it is possible to purify silicon to remove the admixtures forming volatile halogenides:

$$3\ SiCl_4 + Al_4Si_3 \rightarrow 4\ AlCl_3 + 6\ Si \tag{48}$$

The separation of compounds with different volatility may be carried out efficiently by the continuous counterflow process, in which the operations of distillation and condensation are repeated many times.

This process is referred to as rectification which takes place in the rectification column. The rectification column is a continuously or periodically operating system used for the separation of two- or more component mixtures. Rectification and distillation are based on the different volatility of components, and theoretically it is possible to obtain any degree of separation with the result close to 100%. Distillation can be efficiently for separating two elements of the liquid phase if their boiling points differ by more than 50 °C. Rectification makes it possible to separate elements with the boiling points differing by no more than 0.5 °C. At present, rectification is one of the most widely used methods of distillation separation of pure components. The method is relatively simple, equipment is compact, the process is continuous. Because of all these factors, rectification is a widely used method in industry in the production of pure metals and semiconductor materials.

To ensure that the separated mixture can be subjected to rectification, the components of the mixture must fulfil a number of conditions;

- they must be stable in the rectification conditions
- the boiling point of these components must not be too high;
- if possible, they should be characterised by high relative volatility;
- they should not form an azeotropic mixture together;
- if possible, the difference between the boiling and melting points of the components should be large.

The last requirement is very important because to ensure efficient rectification, it is necessary to condense part of the vapours leaving the column and return them back the form of a melt. This is difficult to realise if the melting and boiling points are similar and it is completely impossible if the substance does not melt but is directly transferred to the vapours, i.e. sublimates. Certain elements, Hg, Zn, Cd, Li, Br, can be rectified directly in the main form. However, in a number of cases, the process can be carried out only using volatile compounds. At present, rectification is used in industry in refining of $TiCl_4$, $SiCl_4$, $GeCl_4$, for the separation of compounds of Zr and Hf and for purification of a number of metals.

7.6. Principle of the rectification process

If the flow of the vapours leaving the evaporating surface in distillation maintains the direction of movement identical up to the moment of reaching the condensation surface, rectification is based on the fact that the flow of the liquid – condensed vapours (condenser)-flows against the advancing flow of the vapours. Distillation is accompanied

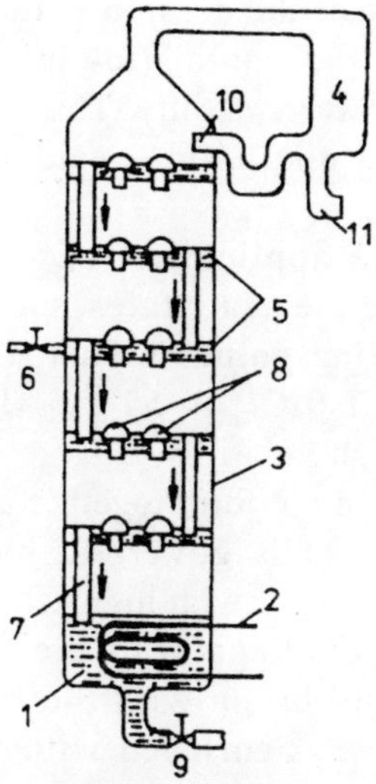

Fig. 46. Rectification column.

only by evaporation and condensation, whereas rectification is characterised, as a result of close contact of both phases, by the transfer of heat and mass and, consequently, the maximum contact surface of these media is important, Fig. 46.

Figure 46 shows the diagram of the rectification column consisting of four main components: 1 boiler, 2 heating element, 3 column, 4 condenser. The rectification column has a number of horizontal barriers 5 in the form of trays in five or more distillation levels. The solution to be rectified is heated and supplied through the gate 6 to one of the central levels, fills the level and flows down through a pipe into a tray on a lower level. On this level, the solution meets the rising vapour which bubbles through the solution, passes through the pipes 8, close with lids ensuring efficient contact between the vapours and the liquid. In contact with the liquid, the less volatile component transfers its evaporation enthalpy to the liquid and condenses to the liquid. Consequently, the content of the component with the boiling temperature higher than that of the vapour in the condenser decreases. At the same time, the liquid receives part of the heat and forms vapours characterised by a higher content of the component with the lower boiling component that of the liquid. Consequently, the vapour passing through the pipe 8 into higher levels is enriched with the more volatile component that the liquid flowing into the lower levels through the pipe 7. This process is repeated on each level and, consequently, in the column with a large number of levels and with efficient regulation of the operating conditions of the column vapours leaving the upper part of the column contain an almost pure more volatile component and the liquid, flowing into the boiler, is formed by a substance with a lower volatility. This liquid is represented by, for example, the so-called distillation residue discharged from the boiler through the gate 9.

A suitable example of the application of rectification is the purification of $TiCl_4$ to remove the admixtures, mostly chlorides with a low boiling point. The boiling point of $TiCl_4$ is 136 °C and, for example, the boiling point of $SiCl_4$ is 57 °C. Tetrachlorides are characterised by sufficiently high volatility, a large difference between the boiling and melting points and not form together azeotropic mixtures, because the $TiCl_4$–$SiCl_4$ system is governed by the Raoult law. Consequently, rectification can be used with high efficiency. As indicated by the temperature–composition diagram of the $TiCl_4$-$SiCl_4$ system (Fig. 47), the vapour is in equilibrium with the liquid; at the given temperature, the vapour is always enriched with low-boiling $SiCl_4$. The first stage of rectification takes place in the conditions ensuring the separation of $SiCl_4$ from $TiCl_4$ which forms the distillation residue. For the separation of other admixtures, i.e. chlorides less volatile than $TiCl_4$, it is necessary to repeat rectification. Another example of the application of rectification is the well-known separation of zinc (t_b^{Zn} = 906 °C) from cadmium with high volatility (t_b^{Cd}= 768 °C).

The theory of rectification of, in particular, two-component mixtures has been developed quite sufficiently and it is supported by a large number of analytical, graphic and calculation methods. The applicability of rectification depends basically on the following factors:

- the distribution number b_R which is typical of the given solution and depends on the composition of the system (concentration of the admixture);
- the number of levels in the column;
- the efficiency of operation of every level of the column, which is associated with the rate of the process and design of the individual levels (trays).

In the rectification of multicomponent mixtures, it is necessary to use several columns after each other. The products are taken from the top of the rectification column (distillate) and from the bottom of the rectification column (distillation residue) or from a certain height (lateral removal).

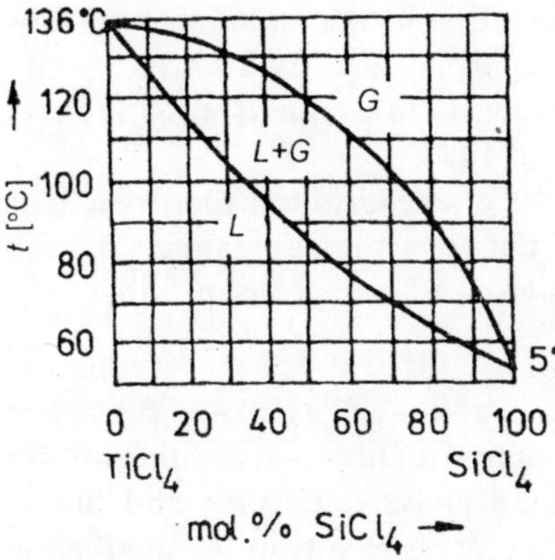

Fig. 47. Composition–temperature diagram of the $TiCl_4$–$SiCl_4$ system.

7.7. Using rectification for the refining and separation of metals and compounds

Several examples will be discussed in which rectification can be used for the efficient separation or purification of metals. Because of their high volatility, the alkaline metals, Li, Na, Rb, Cs, can be efficiently rectified as a metallic melt. This is also possible in the case of Zn, Cd, Hg. The most efficient method of separation of Zr and Hf is the rectification of their tetrachlorides. The efficient rectification of Ta and Nb may be carried out using their fluorides, chlorides, bromides and also tetrachlorides. The purification of V may be carried out efficiently using $VOCl_3$. In the case of W and Mo, it is WCl_5, WCl_6, $WOCl_4$ and $MoCl_5$.

The rectification process may be used efficiently for the refining of compounds of many elements used in semiconductor technology, because many of these elements form volatile halogenides with good prerequisites for rectification. These halogenides are the initial products for obtaining the required elements. In addition to the chlorides of Ge and Si, it is also possible to obtain high purity in the case of Ga, B, Sb by rectification of their halogenides. It is also possible to carry out rectification of Se and Te either as the main components or $SeOCl_2$, $TeCl_2$, $TeCl_4$.

7.8. Distillation by means of chemical transport reactions

Distillation of substances by means of chemical transport reactions as a method for their separation and refining is carried out on the basis of transformation of the main metal into a chemical compound characterised by higher volatility than that of admixtures or other components. In the subsequent dissociation of a compound with low volatility, it is possible to produce distilled metal which is usually characterised by a considerably higher purity in comparison with direct distillation, if this can be carried out [43].

The principle of distillation by means of chemical transport reactions is based on the reversible reaction between the substance (A) and the gas or vapour (B) with the formation of a gaseous product (*C*):

$$mA_{[S,L]} + B_{[G]} \rightleftarrows rC_{[G]} \tag{49}$$

If in one area of equipment this reaction takes place from left to right and in the other area in the opposite direction, in the former part the amount of the solid or liquid phase decreases and in the upper part of equipment increases. This process results in the distillation or transfer of the given substance from one part of equipment to another by means of a chemical transport reaction.

7.9. Distillation of metals by means of a reversible endothermic exchange reaction

The metals and elements characterised by the very low vapour pressure at temperatures used in practice, may be distilled by means of a reversible exchange reaction of the following type:

$$m\,Me_{[cond]} + n\,Ma\,X_{m[G]} \underset{T_2}{\overset{T_1}{\rightleftarrows}} m\,MeX_{n[G]} + n\,Ma_{[G]} \tag{50}$$

where Me is the non-volatile metal with valency n; Ma is the volatile metal with valency m; X is a halogenide.

The distillation of metals by means of the so-called indirect distillation makes it possible to reduce greatly amount of heat required for the transfer of metals from the solid to gas state. The principle of this method is therefore based on the combination of the strongly endothermic process of direct evaporation and the exothermic reaction. Schematically, these reactions may be described by the following equations:

$$m\,Me_{[cond]} \rightleftarrows m\,Me_{[G]} - m\,\Delta H_{ev} \tag{51}$$

$$n\,Me_{[G]} + n\,Ma\,X_{m[G]} \rightleftarrows m\,MeX_{n[G]} + n\,Ma_{[G]} + \Delta H_b \tag{52}$$

where ΔH_{ev} is the change of the enthalpy of evaporation of the metal; ΔH_b in the change of the enthalpy of the exothermic reaction.

The joint occurrence of these reactions results in the transfer of metal Me into a gas compound in the form of MeX;

$$m\,\mathrm{Me}_{[\mathrm{cond}]} + n\,\mathrm{Ma\,X}_{m[G]} \underset{T_2}{\overset{T_1}{\rightleftarrows}} m\,\mathrm{Me\,X}_{n[G]} + n\,\mathrm{Ma}_{[G]} - (m\,\Delta H_{\mathrm{ev}} - \Delta H_b) \quad (53)$$

Therefore, this process is less endothermic than direct evaporation and, consequently, it takes place at a considerably lower temperature. This process is a reversible and, consequently, after transport in cooling to temperature T_2 the re-condensed non-volatile metal (Me) and the vapours of the volatile halogenide of the metal (MaX) form again in a new area as the product of the reaction. In order to ensure that the reaction is reversible, both the metals and halogenides must not form mutually stable compounds in the conditions of the reaction taking place.

The distillation of metals by their stable halogenides is realised in practice by passing the vapours of the halogenides of the volatile metal above the non-volatile metal. The resultant transport gas mixture cools down in a different area of the apparatus and this results in a reverse process and the non-volatile metal condenses completely or partially free from halogenides or contains a certain amount of these halogenides (depending on the properties of the reacting compounds and the process conditions), Fig. 48.

Since the reaction of indirect distillation is endothermic, increase of temperature supports the occurrence of this reaction, and a decrease of pressure increases the reaction rate. The efficiency of separation and refining of metals in indirect distillation depends on the affinity of the metal and individual admixtures for the individual halogenides, i.e. on the decrease of the Gibbs energy with the formation of halogenides and on the concentration of each element in the alloy. A suitable example is the refining of beryllium, titanium and nickel:

$$\mathrm{Be}_{[S]} + 2\,\mathrm{NaCl}_{[G]} \rightleftarrows \mathrm{BeCl}_{2[G]} + 2\,\mathrm{Na}_{[G]} \quad (54)$$

$$\mathrm{Ti}_{[S]} + 4\,\mathrm{NaCl}_{[G]} \rightleftarrows \mathrm{TiCl}_{4[G]} + 4\,\mathrm{Na}_{[G]} \quad (55)$$

$$\mathrm{Ni}_{[S]} + 2\,\mathrm{HCl}_{[G]} \rightleftarrows \mathrm{NiCl}_{2[G]} + \mathrm{H}_{2[G]} \quad (56)$$

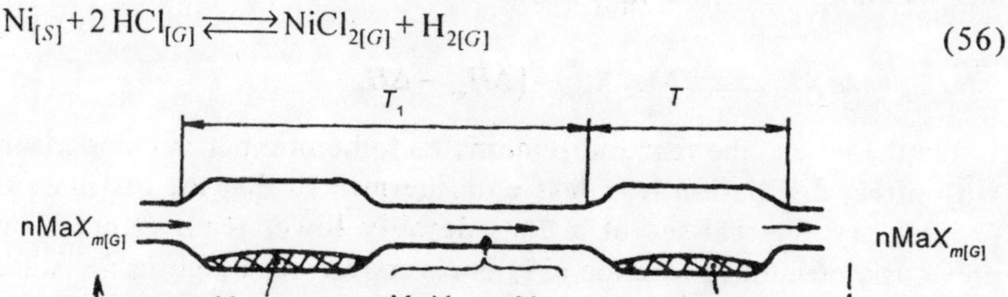

Fig. 48 Distilation of metal [Me] by a reversible endothermic reaction ($T_1 > T_2$): $m\mathrm{Me}_{[\mathrm{cond}]} + n\mathrm{Ma}X_{m[G]} \rightleftarrows m\mathrm{Me}X_{n[G]} + n\mathrm{Ma}_{[G]}$.

In the vacuum conditions, the activity of these reactions is quite high in the range T_1 at the evaporation temperature and the temperature of condensation of chlorides at 800 °C and in the range T_2 at the temperature of the reaction of 1000 to 1250 °C. It should be stressed that at the reaction temperature T_2 = 1400 °C, the amount of distilled titanium is 500 times larger than the amount which would have been obtained in simple evaporation of metallic titanium in the same temperature conditions.

7.10. Distillation of metals through sub-compounds

Another method of obtaining a high degree of purification and separation of metals by means of chemical transport reactions is distillation or sublimation of sub-compounds of metals. The sub-compounds are highly volatile non-saturated compounds of metals with a reduced valency. The compounds with reduced valency form in cases in which not all valency electrons of the atom take part in the chemical reaction, and only those characterised by the weakest bonding with the core participate, i.e. those characterised by the lowest value of the ionisation potential. The sub-compounds form at higher temperatures, mainly in the inert atmosphere or in vacuum, with the simultaneous reaction of the atom of the metal with some of the chemical compounds of the same metal, for example, halogenide:

$$\mathrm{Me} + \mathrm{MeX} \underset{T_2}{\overset{T_1}{\rightleftarrows}} \mathrm{Me_2X} \tag{57}$$

In this case, the process becomes less endothermic in comparison with simple sublimation or distillation. The reaction may be described as the sum of two reactions:

$$\mathrm{Me_{[cond]}} \rightleftarrows \mathrm{Me}_{[G]} - \Delta H_{ev} \tag{58}$$

$$\mathrm{Me}_{[G]} + \mathrm{MeX}_{[G]} \rightleftarrows \mathrm{Me_2X}_{[G]} + \Delta H_b \tag{59}$$

$$\mathrm{Me_{[cond]}} + \mathrm{MeX}_{[G]} \underset{T_2}{\overset{T_1}{\rightleftarrows}} \mathrm{Me_2X}_{[G]} - (\Delta H_{ev} - \Delta H_b) \tag{60}$$

On the whole, the reaction remains endothermic but in comparison with direct distillation it is less endothermic so that the distillation process may be realised at a considerably lower temperature. The previously mentioned reaction (59) is reversible. At a high temperature (T_1) reaction (60) takes place in the direction of formation of volatile sub-compounds. After cooling to temperature (T_2), the sub-compound formed in this manner dissociates into a metal and its compound with a high valency, Fig. 49. In this case, the carrier of distilled metal

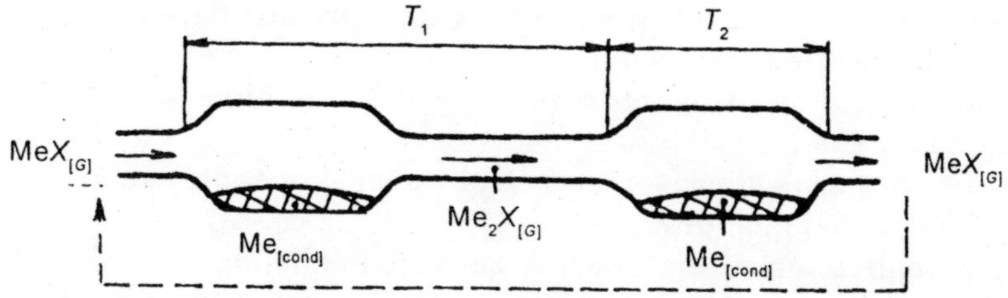

Fig. 49 Diagram of distillation of metal through sub-compounds, transfer of metal to the cold zone, $T_1 > T_2$: $Me_{[cond]} + MeX_{[G]} \rightleftarrows Me_2X_{[G]}$.

is the volatile compound with reduced valency (sub-compound) of the same metal. An increase of temperature and a decrease of pressure accelerate the reaction of formation of the sub-compound and, consequently, the process of transport distillation of the metal.

The formation of compounds of the univalent subchloride AlCl at higher temperatures ($T_1 \approx 1000$ °C) is a consequence of the chemical reaction of metallic aluminium with $AlCl_3$ compound with a higher valency:

$$2\,Al_{[cond]} + AlCl_{3[G]} \underset{T_2}{\overset{T_1}{\rightleftarrows}} 3\,AlCl_{[G]} \tag{61}$$

The compounds of the univalent metal usually exist in the gas state (in the inert atmosphere or in vacuum). With a decrease of temperature ($T_2 \approx 600$ °C), the equilibrium of the reaction of the formation of sub-compounds of metals is displaced the left and these compounds dissociate into the transported metal (aluminium) and its compounds with a higher valency ($AlCl_3$).

The capacity of aluminium to form volatile compounds with a lower valency, in particular, subfluorides and subchlorides, may be utilised for the separation of aluminium from other metals and for deep refining to remove admixtures. The refining of aluminium by means of sub-compounds takes place in special distillation columns in accordance with the reactions:

$$AlF_{3[G]} + 2\,Al_{[L]} \xrightarrow{T_1} 3\,AlF_{[G]} \xrightarrow{T_2} 2\,Al_{[L]} + AlF_{3[G]} \tag{62}$$

$$AlCl_{3[G]} + 2\,Al_{[L]} \xrightarrow{1000\,°C} 3\,AlCl_{[G]} \xrightarrow{600\,°C} 2\,Al_{[S]} + AlCl_{3[G]} \tag{63}$$

Other examples:

$$Si_{[S]} + SiCl_{4[G]} \rightleftarrows 2\,SiCl_{2[G]} \tag{64}$$

$$Ti_{[S]} + TiCl_{4[G]} \rightleftarrows 2\,TiCl_{2[G]} \tag{65}$$

The conditions for obtaining a pure metal by distillation of these metals through subhalogenides are:

- the application of the initial metal with the minimum content of admixtures;
- efficient purification, for example, by double or triple sublimation of initial halogenides;
- the application of high quality corrosion-resisting materials for the manufacture of components of equipment with which the refined metal and its halogenides come into contact.

Table 8 shows examples of the transport reactions by means of chlorides and subchlorides of metals for different types of the chemism of the reactions [43]:

1. Transport of metals by means of sub-compounds
2. Transport of chlorides and oxychlorides of metals
3. Transport of oxides and other compounds.

7.11. Distillation by synthesis and dissociation of volatile compounds. Iodide reaction

In 1925, van Arkel proposed the so-called iodide process for refining of refractory metals.

The principle of the method is based on the following: the volatile iodide of a metal is heated and evaporated in a vessel 6 produced from glass with a higher melting point (Fig. 50) in which the thin W-wire (fibre) 4 is suspended and secured at points 5 and 8 between tungsten bars 1, 2, brazed into the vessel. The passage of electric current through W-wire results in heating of the wire to a high temperature, $t_2 > 1400$ °C, accompanied by the dissociation of the iodide. Consequently, the metal settles on the surface of the wire in pure form. The iodide of the metal is obtained at a different temperature in a different area of the same vessel. The initial powder

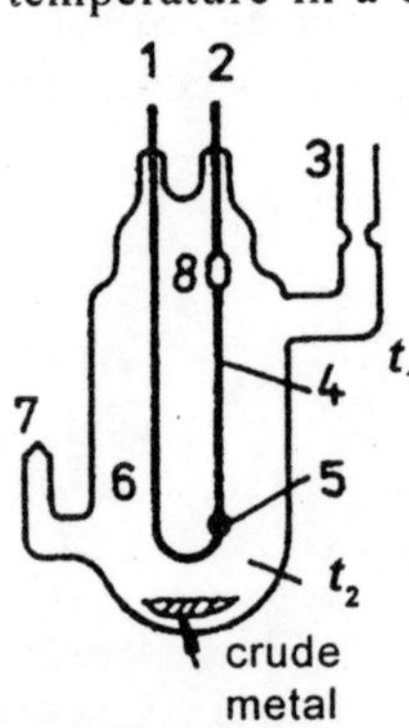

Fig. 50 Diagram of the reaction vessel for the iodide reaction of metals (van Arkel).

metal and the iodide are introduced into the vessel through the pipe 7 which is subsequently sealed by melting. Vacuum is produced in the vessel and the pipe 3 is also sealed by melting. The entire system is heated in an electric furnace to temperature $t_1 > 200$ °C at which the initial metal and iodide reacted to form the metal tetraiodide and evaporate. Temperature t_1 is considerably lower than the temperature to which the tungsten wire is heated and at which thermal dissociation of the iodide takes place. The transport dissociation iodide reaction may be expressed by the following schematic equation:

$$Me_{[S]} + 2\,I_{2[G]} \underset{T_2}{\overset{T_1}{\rightleftarrows}} Me\,I_{4[G]} \tag{66}$$

and $T_1 < T_2$.

Schematically, the process is shown in Fig. 51. In this case, the metal is transported from the colder to warmer zone. The reaction of formation of a volatile intermediate product – carrier (MeI_4), is accompanied by a thermal effect. The thermal dissociation of iodide results in the formation of a number of high-purity metals, for example, Ti, Zr, Ta, Nb, V, and others. As an example, Table 9 shows the temperatures (t_2) of settling of dissociated metals on the wire and temperatures (t_1) of synthesis and evaporation of the iodides of metals.

The iodide method is also used for the refining of metals to remove both interstitial admixtures, oxygen, nitrogen, carbon and also metallic admixtures which form iodides with lower volatility than the main metal. This process can not be used in the case of metals whose melting point is lower than the dissociation temperature of the iodide. This would make it impossible to maintain these metals as a melt on the wire.

From the kinetic viewpoint, the process of iodide refining consists of the following stages:

- Synthesis of the iodide as a result of the reaction of the vapours of iodine with a metal;
- Transfer of the iodide to the heated wire;
- Thermal dissociation of iodide on the wire and deposition of the

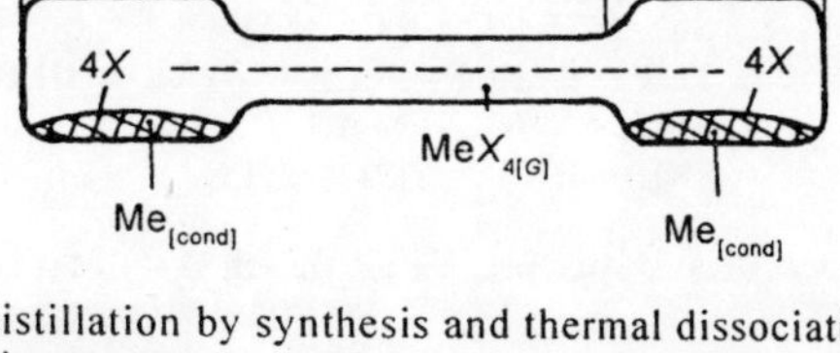

Fig. 51 Diagram of distillation by synthesis and thermal dissociation of metal iodide $Me_{[cond]} + 2I_{2[G]} \underset{T_2}{\overset{T_1}{\rightleftarrows}} MeI_{4[G]}$.

Table 8. Transport reaction of chlorides and subchlorides of metals

Element	Transport compound	Transport reaction	Conditions t_1/t_2 [°C]
		1. Transport of metals by means of sub-compounds	
Be	NaCl	$Be + 2NaCl = BeCl_2 + 2Na$	1000/21
B	BCl_3	$2B + BCl_3 = 3BCl$	
Ga	$GaCl_3$	$2Ga + GaCl_3 = 3GaCl$	
Zr	$ZrCl_4$	$Zr + 3ZrCl_4 = 4\ ZrCl_3$	
Nb	$NbCl_5$	$Nb + 4NbCl_5 = 5NbCl_5$	
Ta	$TaCl_5$	$Ta + 4TaCl_5 = 5TaCl_4$	
Ti	$TiCl_3$	$Ti + 2TiCl_3 = 3TiCl_2$	1200/1000[A]
Mo	Cl_2	$Mo + 5/2Cl_2 = MoCl_5$	1400/400
W	Cl_2	$W + 3Cl_2 = WCl_6$	1400/400
Au	Cl_2	$2Au + 3Cl_2 = Au_2Cl_6$	
Al	$AlCl_3$	$2Al + AlCl_3 = 3AlCl$	1000/600
Si	$SiCl_4$	$Si + SiCl_4 = 2SiCl_2$	
Si	$AlCl_3$	$Si + AlCl_3 = SiCl_2 + AlCl$	1000/21
Fe	HCl	$Fe + 2HCl = FeCl_2 + H_2$	1000/800
Co	HCl	$Co + 2HCl = CoCl_2 + H_2$	900/600
Cu	HCl	$3Cu + 2HCl = CuCl + 1/2H_2$	600/500
		2. Transport of chlorides and oxychlorides of metals	
VCl_3	Cl_2	$VCl_3 + 1/2Cl_2 = VCl_4$	300/250
VOCl	VCl_4	$VOCl + 2VCl_4 = VOCl_3 + 2VCl_3$	800/700
$NbCl_3$	$NbCl_5$	$NbCl_3 + NbCl_5 = 2NbCl_4$	390/355
$NbOCl_2$	$NbCl_5$	$NbOCl_2 + NbCl_5 = NbOCl_3 + NbCl_4$	400/350
$TaCl_3$	$TaCl_5$	$TaCl_3 + TaCl_5 = 2\ TaCl_4$	580/360
TiOCl	HCl	$TiOCl + 2HCl = TiCl_3 + H_2O$	650/550
$MoCl_3$	$MoCl_5$	$MoCl_3 + MoCl_5 = 2MoCl_4$	300/250
AlOCl	$NbCl_5$	$AlOCl + NbCl_5 = 1/2Al_2Cl_6 + NbOCl$	400/380
CrOCl	$CrCl_4$+Cl_2	$2CrOCl + CrCl_4 + Cl_2 = CrO_2Cl_2 + 2CrCl_3$	1000/840
FeOCl	HCl	$FeOCl + 2HCl = FeCl_3 + H_2O$	~350
		3. Transport of oxides and other compounds	
BeO	HCl	$BeO + 2HCl = BeCl_2 + H_2O$	1000/800[D]
Ta_2O_5	Cl_2	$Ta_2O_5 + 3Cl_2 = 2TaOCl_3 + 3/2O_2$	950/700[A]
Ta_2O_5	HCl	$Ta_2O_5 + 6HCl = 2TaOCl_3 + 3H_2O$	1000/600[D]
Ta_2O_5	$TaCl_5$	$Ta_2O_5 + 3TaCl_5 = 5TaOCl_3$	750/650[A]
Al_2O_3	HCl	$Al_2O_3 + 6HCl = 2AlCl_3 + 3H_2O$	1000/21[D]
Al_2O3	Cl_2	$Al_2O_3 + 3Cl_2 = 2AlCl_3 + 3/2O_2$	1240/21[D]
Fe_2O_3	HCl	$Fe_2O_3 + 6HCl = 2FeCl_3 + 3H_2O$	1000/800[D,A]
SiO_2	$TaCl_5$	$SiO_2 + 2TaCl_5 = SiCl_4 + 2TaOCl_3$	1050/21
$FeWO_4$	HCl	$FeWo_4 + 6HCl = FeCl_2 + WOCl_4 + 3H_2O$	
MoS_2	Cl_2	$MoS_2 + 7/2Cl_2 = MoCl_5 + S_2Cl_2$	950/800
TiO_2	HCl	$TiO_2 + 4HCl_4 = TiCl_4 + 2H_2O$	1000/21

Comment: A – process takes place in a sealed vessel; D – metal forms on a heated wire.

Table 9 Pure metals produced by thermal dissociation of their iodides

Element	Reaction temperature in vessel t_1 [° C]	Dissociation temperature on wire t_2 [°C]	Element	Reaction temperature in vessel t_1 [° C]	Dissociation temperature on wire t_2 [°C]
Be	400	800	Si	200	1000
Cr	800	1100	Ta	500	1100
Cu	400	1100	Ti	200	1400
Hf	800	1600	Cr	850	1150
Nb	360	900	V	900	1300
Fe	500	1100	Zr	200	1400

metal on the wire;

- Transfer of released iodine from the wire back to the iodine container.

The kinetics of association of volatile iodide, i.e. the rate of deposition of metal, is affected by the following factors:

- Amount of iodine in the vessel;
- Reaction temperature in the vessel;
- Temperature of heated wire;
- Formation of lower iodides.

The metals which can be efficiently refined by the van Arkel method, in which all the previously mentioned reactions take place in the single reaction vessel, include Ti, Zr, Hf. At temperatures at which the refining process is realised by volatile iodides, various compounds of the metals to be refined, such as nitrides, carbonitrides, silicides and also other intermetallic compounds, such as oxides, carbides, do not take part in these iodide transport reactions. The iodides of the bivalent metals, if they form, are characterised by insignificant vapour pressure and, consequently, do not take part in the transfer process.

The iodide process can also be used for the production of alloys, if the reaction vessel is filled with a powder mixture of metals. For example, this method was used to produce alloys: Zr–Sn, Zr–Al, Ti-Zr, Zr–Hf, Zr–Cr, Zr–Ni, with a different concentration of the components.

7.12. Suitability of the iodide process for different groups of metals

Below, there is a brief survey of the suitability of using the iodide process for different groups of metals in accordance with the periodic table of elements. For the metals of the group I and II there are no volatile compounds which would dissociate visibly below the melting point of the metals. An exception from the elements of group II is beryllium which settles if the metallic fibre is heated in the vapours of BeI_2. Of the elements of group III, more suitable conditions for production of boron exist in its reaction with bromide. The group III also includes lanthanides. Their melting points are relatively high and, consequently, it may be possible to produce metals from these iodides. This method has been used for the preparation of yttrium. The elements of group IV (Si, Ti, Zr, Hf) are normally produced by dissociation of their iodides. The metals produced by this method are characterised by high ductility and good machinability. From the metals of group V (V, Nb, Ta), it is possible to produce only a limited number of compounds, for example, vanadium from iodides. However, since the vanadium tri-iodide VI_3 is unstable, it is evident that the vanadium dioxide forms, which like a compound is characterised by low volatility, like a series of bivalent compounds of the type MeI_2 and, consequently, the quartz reaction vessel must be heated to 800 °C. More favourable conditions exist in the case of Nb and Ta which can be produced not only by thermal dissociation of the iodides but also chlorides. Chromium can be produced using a similar method as in the case of vanadium. Of the elements of group VII, it has not as yet been possible to prepare manganese by this method, because the iodide of this element is characterised by low volatility, but on the other hand Re, as a refractory metal, can be produced by the dissociation of its chlorides. Of the elements of the group VIII, in particular the elements of the iron group, only iron can be produced by dissociation of iodide like iron. So far, nickel and cobalt could not be refined by the iodide method. This may be explained by the fact that these metals probably do not form iodides and their di-iodides have low volatility.

In practice, nickel of relatively high purity is obtained by dissociation of nickel tetracarbonyl $Ni(CO)_4$

$$Ni + 4\,CO \underset{200\,°C}{\overset{60\,°C}{\rightleftarrows}} Ni(CO)_4 \qquad (67)$$

The carbonyl method is also suitable for the preparation of iron powder of only commercial purity from pentacarbonyl $Fe(CO)_5$, contaminated of course with carbon, oxygen and traces of admixtures of other metals, capable of forming carbonyl.

The iodide process may be used efficiently for the deposition of very thin layers with a thickness of up to 10 μm on the substrate of various semiconductor materials. The efficiently purified iodide of a specific element is subjected to thermal dissociation on the surface of the substrate, heated to a suitable reaction temperature. This results in the formation of single crystal layers so that the technology can be used for the preparation of miniature semiconductor components applied in radio engineering, in computers and automation systems.

A classic example is at present the deposition of single crystal layers on silicon. The efficiently purified tetrachloride $SiCl_4$ or $SiHCl_3$ is reduced by hydrogen on the surface of the substrate, heated to 1000–1200 °C:

$$SiCl_4 + 2\,H_2 \rightarrow Si + 4\,HCl \tag{68}$$

$$SiHCl_3 + H_2 \rightarrow Si + 3\,HCl \tag{69}$$

Reaction (69) requires a small amount of hydrogen for reduction. Similarly, in the metallurgy of semiconductors it is possible to deposit on substrates of more complicated shapes gradually several different layers with different semiconductor, semi-insulating or dielectric properties.

8. ELECTROLYSIS

8.1. Significance of electrolytic method of separation and refining of metals

Electrolysis is one of the efficient and widely used methods of separation and refining of metals. The advantage of this method in comparison with other methods is that it makes it possible to carry out the process on large scale and almost continuously [45]. In addition to this, the application of electrolytic method of separation and refining of metals is not limited by the temperatures of sublimation or boiling point of metals. Electrolysis may be carried out in water and water-free solutions, in salt melts, using solid or liquid electrodes. Pure metals can be produced by electrolysis either by means of insoluble anodes (electrolytic precipitation) or soluble anodes (electrolytic refining).

The electrolytically produced and refined metals can be used in a number of cases directly as metallic cathodes in various areas of technology. Electrolysis is used in most cases only for the first stage of purification and the resultant cathode metal is the initial material for its further deep refining, for example, by vacuum remelting, zone melting, distillation by means of chemical transport reactions, etc. At present, electrolytic precipitation and refining from aqueous solutions is used for the preparation of a number of pure metals, such as copper, nickel, zinc, lead, indium, gallium, noble metals silver, gold. Electrolysis of salt melts of oxygen-free acids is used to produce aluminium, magnesium and other light metals.

8.2. The rate of the electrode process. Electrochemical and diffusion kinetics

For general evaluation of the sequence of precipitation of cations on the cathode or the transfer of atoms to the ion condition on the anode it is necessary to use the so-called electrochemical voltage series, i.e. the values of electrode potentials obtained in the equilibrium conditions, i.e. the standard potentials. The electrolytic voltage series

(Table 10) shows that: as the cation becomes more positive, the probability of its discharge on the cathode increases and the probability of anodic dissolution of the corresponding metal decreases.

In both anode dissolution and in the precipitation of metal on the cathode, it is not the elementary act but in fact electrode processes representing complicated electrochemical phenomena. The electrode process contains all changes which take place if the potential differs from the equilibrium potential.

The individual stages of the electrode process are:

- the transfer of the ion from the volume of the solution to the surface of the electrode to the distance at which the electrochemical reactions, reduction or oxidation of the ion may take place. In most cases, this stage takes place at the expense of diffusion which occurs as a result of the difference of the concentration of the dissolved substance in the volume of the solution and on the surface of the electrode. Here the concentration of the ion of the dissolved substance is lower because the substance is being used up during the electrochemical reaction. Without continuous supply of the ions to the surface of the electrode, the electrode process cannot take place.
- the actual electrochemical reaction in which the ions of the substance change their charge. On the cathode, the electrons transfer from the electrode to the ion and reduce this ion; on the anode, the ion of the substance transfers its charge and oxidises.
- the formation of a new substance, the final product of electrolysis. If this product is a gas (for example, hydrogen), the final stage is the formation of molecules from atoms, the formation of bubbles and their removal from the surface of the electrode. If the product of electrolysis is a solid substance (a reduced metal), the final stage is characterised by the formation of the crystal of the metal. The simultaneous discharge of two or more different cations may lead to the formation of a solid solution of the alloy or an intermetallic compound. Finally, if the product of the process is the ion remaining in the solution (for example, in the reduction of Fe^{3+} to Fe^{2+}) the final stage consists of the displacement of this product from the area of formation, i.e. from the surface of the electrode to the entire volume of the solution.

Depending on which stage is characterised by the lowest rate, there is the electrochemical and diffusion kinetics of the electrode process characterised by different relationships.

The electrolytic precipitation of metals is quantitatively determined by Faraday laws. These laws specify the relationship between the

Table 10 Electrochemical voltage series of elements (standard potentials E^0 at 25 °C, external pressure 101325 Pa)

Element/ion	E^0 [V]	Element/ion	E^0 [V]	Element/ion	E^0 [V]
Li/Li^+	–3.045	Cr/Cr^{2+}	–0.560	As/As^{3+}	+0.247
Rb/Rb^+	–2.926	Ga/Ga^{3+}	–0.530	Cu/Cu^{2+}	+0.337
K/K^+	–2.925	Fe/Fe^{2+}	–0.440	Co/Co^{3+}	+0.430
Cs/Cs^+	–2.920	Cd/Cd^{2+}	–0.403	Cu/Cu^+	+0.521
Ba/Ba^{2+}	–2.900	In/In^{3+}	–0.340	$I_2/2I^-$	+0.535
Sr/Sr^{2+}	–2.888	Te/Te^+	–0.336	Rh/Rh^+	+0.600
Ca/Ca^{2+}	–2.866	Co/Co^{2+}	–0.277	Te/Te^{3+}	+0.720
Na/Na^+	–2.714	In/In^+	–0.250	$2Hg/Hg_2^{2+}$	+0.788
La/La^{3+}	–2.522	Ni/Ni^{2+}	–0.250	Ag/Ag^+	+0.799
Mg/Mg^{2+}	–2.370	Sn/Sn^{4+}	–0.136	Rh/Rh^{3+}	+0.800
Be/Be^{2+}	–1.847	Pb/Pb^{2+}	–0.126	Hg/Hg^{2+}	+0.854
Ti/Ti^{2+}	–1.750	Fe/Fe^{3+}	–0.036	Pd/Pd^{2+}	+0.987
Al/Al^{3+}	–1.662			$Br_2/2Br^-$	+1.065
Mn/Mn^{2+}	–1.180	**H_2/H^+**	**0.000**	Pt/Pt^{2+}	+1.200
Te/Te^{2-}	–1.140			$Cl_2/2Cl^-$	+1.360
Se/Se^{2-}	–0.920	Sn/Sn^{4+}	+0.010	Au/Au^{3+}	+1.500
Zn/Zn^{2+}	–0.763	Sb/Sb^{3+}	+0.212	Au/Au^+	+1.691
Cr/Cr^{3+}	–0.744	Bi/Bi^{3+}	+0.226	$F_2/2F^-$	+2.850

electric charge transferred through the electrolyte and the chemical effect caused by this charge. They may be formulated as follows:

– the amount of substance consumed or formed on the electrode is proportional to the transmitted charge:

$$m = q\, I\, \tau \tag{70}$$

where m is the amount of the precipitated substance [g]; q is the electrochemical equivalent of the substance precipitated by the passage of a charge of 1 coulomb [g C^{-1}]; I is the electric current [A], τ is the electrolysis time [s];

– the amounts of substances used up or formed in different electrochemical phenomena during the passage of the same electric charge are proportional to the electrochemical equivalent of these substances:

$$m = \frac{M_A\, I \tau}{z F} \tag{71}$$

where M_A is the atomic mass of the precipitated elements; z is the number of exchange electrons (valency of the substance); F is the Faraday constant (F = 96487+16 C mol^{-1} = 26.8 A h mol^{-1}).

For example, a current of 1 A results in the precipitation of

0.5 g H_2, 8 g O_2, 63.54 g Cu in 96487 seconds, or in the dissolution of 65.4 g Zn, 107.87 g Ag.

The rate of the electrochemical reactions is controlled by the exponential law. For the cathodic reaction

$$K^{z+}k + z_c e \rightarrow K \quad (72)$$

the equation for the rate of the electrochemical reaction has the following form

$$I_c = K_1 a_c \exp\left(-\frac{\alpha z_c F \varphi}{RT}\right) \quad (73)$$

For the anodic reaction

$$A^{z-}a - z_a e \rightarrow A \quad (74)$$

the equation for the rate of the electrochemical reaction is:

$$I_a = K_2 a_a \exp\left(-\frac{\beta z_a F \varphi}{RT}\right) \quad (75)$$

where I_c, I_a are the rates of the electrochemical reactions, expressing current density [A cm^{-2}]; α, β are other kinetics factors, whose values are in the range 0–1, and for which $\alpha+\beta = 1$; z_c, z_a is the valency of the cation and the anion; K_1, K_2 are constants of the electrochemical reactions; φ is the potential [V]; a_c, a_a are the activities of the reacting substances.

The rate of the diffusion process can be expressed by the following relationship:

$$I_c = K_d(a - a_s) \quad (76)$$

where K_d is the rate constant of convective diffusion; a is the activity of diffusing particles in the volume of the solution; a_s is the activity of diffusing particles of the surface of the cathode.

The identical equation (76) may also be written for the anode. The rate of diffusion depends on:

- the temperature
- the nature of the dissolved substance and the solvent;
- the intensity of movement of the solution around the electrode, which in the case of the cathode is expressed by constant K_d.

This shows that if the rate of mixing of the solution is changed, i.e. the rate of diffusion, the same electrode process can be transferred from the region of diffusion kinetics to the electrochemical region and vice versa.

8.3. Polarisation curves of reduction of cations

The rate of the electrode process is determined by the rate of the electrochemical reaction only if the concentration of the reacting ions at the surface of the electrode is maintained constant at the expense of the rate of diffusion from the depth of the volume of the solution or from the surface of the electrode to the depth of the volume of the solution. The rate of diffusion must be so high that the rate of the reaction cannot change because of the change in the concentration of the reacting ions. In this case, the rate of the chemical reaction is so low that it determines the overall kinetics of the electrode process.

However, this situation may arise in the case of a relatively low current densities when the consumption of the ions oxidised or reduced on the electrode is low and diffusion is sufficient for supplying the reacting substances in the surface and for removing the reaction products. However, at a high current density, the rate of the electrochemical reaction increases so much that the diffusion is no longer capable of ensuring the supply of ions to the electrode in the required amount and the reaction rate is limited. Now, the diffusion rate is considerably lower than the rate of electrochemical reduction or oxidation and the kinetics of the electrode process is limited by the rate of diffusion. If the electrode reaction enters the diffusion range, any further change of the potential is not capable of increasing the rate of the process: diffusion is not capable of supplying the electrode with the reacting substances, ions.

The polarisation curve, describing the rate of reduction of the cations I_c in relation to the potential, and having the exponential form at low current densities, deviates from this law at higher values of φ and the value of current I_c, i.e. the rate of reduction of the cations, reaches a certain maximum value, determined by the rate of supply of the ions to the surface of the electrode, i.e. by the diffusion rate. This maximum value of current is referred to as the threshold current density, Fig. 52.

Because in this case attention is given to the cathode process, the density of cathode current I_c is situated above the origin of the coordinates. On the right-hand side of the horizontal axis there is the change of the potential, and the line *aa* approximately defines the range of electrochemical kinetics where the exponential relationship between I_c and φ is valid (73). The line *bb* indicates the region of purely diffusion kinetics. The region between the lines *aa* and *bb* is the region of mixed kinetics in which the rate of the electrode reaction and the diffusion rate are comparable. In the case of small deviations from the equilibrium potential, the process is controlled by the electrochemical kinetics (Fig. 52): with increase of cathodic polarisation, the reaction rate increases and the diffusion rate starts to decrease because it does not manage to supply the sufficient number of ions per unit time to the electrode surface. With an increase of the degree of polarisation this rate reduction becomes greater and, consequently, the rate of the electrode process is more and more limited by diffusion. Finally, there are suitable conditions for the threshold current density $I_{p(c)}$ where the diffusion rate has already reached the maximum possible value. Here the curve is parallel to the horizontal axis which indicates that it is not possible to change the reaction rate by increasing the polarisation. In these conditions:

$$\frac{dI_c}{d\varphi} = 0 \tag{77}$$

If the solution contains only one type of cations capable of reduction on the cathode, a further increase of the degree of polarisation cannot cause any other electrochemical process up to the value of the polarisation voltage at which hydrogen starts to generate at the expense of dissociation of water molecules. If the solution contains salts of two different metals whose ions can reduce on the cathode, the polarisation curve is more complicated, Fig. 53a. In accordance with the nature of the metals M_I and M_{II} and the activity of their ions in the solution, the metals will have different equilibrium potentials, indicated on the graph. It may be seen that the potential of M_I is more positive than that of M_{II}.

With a further increase of the degree of cathodic polarisation of the electrode, the equilibrium potential φ_{eq} is reached initially. When the gradually increasing potential of the electrode is more negative than equilibrium potential for M_I, the reduction of the cations M_I takes place and the current I_c, indicating the rate of reduction will start to flow. In the case of a low degree of polarisation, the process is controlled by electrochemical kinetics. With an increase of the degree of polarisation, starting from a specific potential, the rate of the process is inhibited by diffusion. The curve, expressing the law of electrochemical kinetics (broken curve I), and the observed polarisation curve (solid line) separate. Finally, at a specific degree of polarisation, the conditions of threshold current density $I_{p(I)}$ form for the cations M_I. All this takes place at the potentials more negative than $\varphi_{eq(I)}$ but more positive than $\varphi_{eq(II)}$. Therefore, the M_{II}^{2+} ions do not take part in the reduction.

When the electrode potential becomes more negative than the

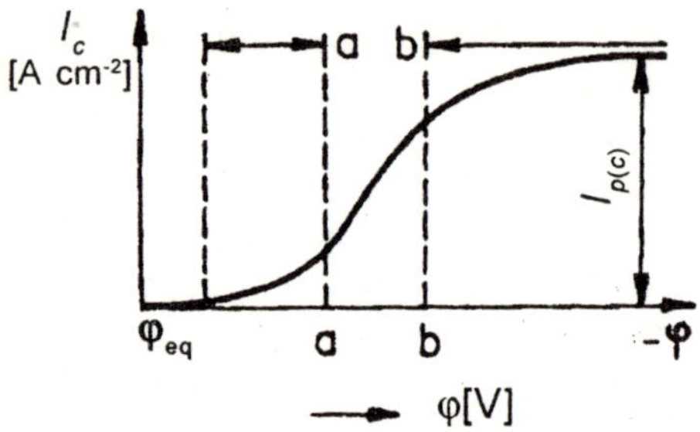

Fig. 52 Cathodic polarisation curve characterised by the electrochemical diffusion kinetics.

equilibrium value for the ions M_{II}^{2+}, the ions start to reduce. At a potential only slightly more negative than $\varphi_{eq(II)}$, the rate of reduction of the ions M_{II}^{2+} will be controlled by the electrochemical kinetics. The layer, adjacent to the cathode, depleted to the maximum extent in ions M_I^{2+} is rich in ions M_{II}^{2+} and the reaction $M_{II}^{2+} \rightarrow M_{II}$ develops without obstacles. At this potential, both types of cations are reduced on the electrode: the ions M_I^{2+} at the limiting rate, determined by the values of $I_{p(I)}$ and the ions M_{II}^{2+}, with the rate given by the laws of electrochemical kinetics. This law is indicated in Fig. 53a by the broken line II. The observed current density represents the sum of the coordinates of the curves II and $I_{p(I)}$. Therefore, if the potential increases to more negative values above $\varphi_{eq(II)}$, the current density increases. A further increase of the negative polarisation results in the same changes of the reduction rate M_{II}^{2+} as those described previously for M_I^{2+}. Finally, suitable conditions form for the threshold current density also for the ions M_{II}^{2+}, $I_{p(II)}$. In this case, the density of the current on the cathode is:

$$I_{p(c)} = I_{p(I)} + I_{p(II)} \tag{78}$$

The two types of ions capable of reduction produced two waves on the polarisation curve, with both the waves changing to a platform, corresponding to the threshold current density of these ions. If the solution contains more types of ions, more waves appear on the polarisation curve.

The above considerations indicate that the simultaneous discharge of two or more ions is possible if the cathode potential is more negative than the standard potential for these processes in the given specific conditions. The density of cathodic current indicating the rate of the reaction, will be equal to the sum of the current densities of all parallel cathodic reactions.

8.4. Simultaneous discharge of the ions of the main metal and admixtures

Up to now, polarisation curves were used to examine the simultaneous discharge of two or more cations whose concentration in the solution/electrolyte was the same. When discussing the possibility of transfer of the admixtures from the electrolyte into the cathodic precipitate, the concentration of their ions in the solution is many times smaller than the concentration of the ions of the main metal. Therefore, the admixtures may be more electropositive and also more electronegative than the main metal. This case of the simultaneous discharge of the ions of the main metal and the admixtures will now be discussed.

It is evident that the discharge of the ions of the main metal and admixtures always takes place simultaneously on the surface of the electrode at the given total potential of the cathode. In the electrolysis conditions in the production of pure metals, when the concentration of the admixtures in the cathodic precipitate is very low, almost all metal admixtures form either diluted solid solutions of the admixtures in the main metal or solid solutions of intermetallic compounds.

The discharge of the ions of the admixtures also takes place if the concentration in the solution is low. If we consider the high density of cathodic current used in experiments, it may be expected that the ions of many admixtures precipitate in the conditions of technical electrolysis or beyond the threshold current density. The electrolysis of the water solutions is usually carried out at a current density of approximately 500 A m^{-2} and the threshold current density, for example, for the sulphate solutions of copper at admixture concentrations in the solution experienced in practice, it is several units of A m^{-2}.

This question will be examined in greater detail by means of polarisation curves. It is assumed that the polarisation curves of the

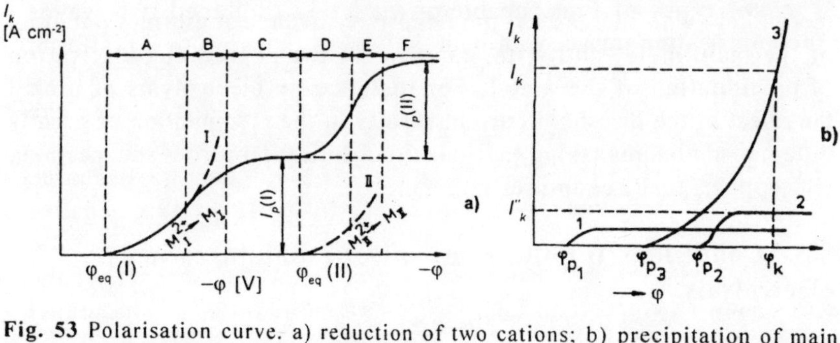

Fig. 53 Polarisation curve. a) reduction of two cations; b) precipitation of main metal and admixtures.

main metal (curve 3) and of two different admixtures (curves 1 and 2) are situated in the manner shown in Fig. 53b. It is assumed that the standard potentials of both admixtures are φ_{p1} and φ_{p2}. The potential of the cathode at the given moment at the current density I_c is denoted by φ_c.

At a sufficiently high concentration of the ions of the first admixture in the solution and with an increase of the cathodic potential, there would be only be the discharge of the ions of the admixture in accordance with curve 1, but since the concentration of the ions of the admixture in the solution is very low, the threshold current is already obtained at a low current density and a horizontal section appears on the polarisation curve. The cathode potential increases without any increase of the current density up to the start of precipitation of the main metal. A further increase of the cathode potential already leads to the simultaneous precipitation of the main metal and the first admixture. The overall polarisation curve may not be identical with curve 3 and must be displaced upwards by the value of the threshold current density of the first admixture in parallel with the curve 3. However, since this current density is low, it may be assumed, without causing any error, that the total polarisation curve is identical with curve 3. The second admixture starts to precipitate together with the main metal and the first admixture starts to precipitate only after reaching current density I_c and, subsequently, the admixtures precipitate at the threshold current density. Thus, both the admixtures with the more positive and negative equilibrium potential may precipitate at the threshold current.

The degree of contamination of the metal within the admixtures precipitated under the effect of threshold current, is directly proportional to the concentration of the admixtures in the solution and indirectly proportional to the cathodic current density of the main metal. The precipitation of the admixtures under effect of threshold current becomes more likely with increase in the negativity of the equilibrium potential of the main metal and with increase in the degree of polarisation of precipitation of the metal. For instance, in electrolysis of nickel the effect of the threshold current results in the precipitation of greatly differing admixtures taking their electrochemical properties into account, e.g. copper, lead, cadmium and zinc.

8.5. Conditions for the removal of admixtures in electrolysis

As mentioned previously, the rate of precipitation of the main metal and the admixtures on the cathode is determined by:

- the rate of supply of the ions to the surface of the cathode as a result of convective diffusion;
- the rate of actual discharge of the ions and precipitation of the metal.

In the conditions in which the precipitation of the main metal is accompanied by the precipitation of admixtures in low concentrations, these admixtures form usually substitutional or interstitial solid solutions. In the majority of cases, the formation of solid solutions leads to extensive depolarisation leading to more efficient precipitation of many admixtures with considerably higher electronegativity in comparison with the main metal. This is limited only by the rate of convective diffusion.

Depending on the stages controlling the rate of precipitation of the admixtures and the main metal, there are four main types of relationship governing the simultaneous discharge of their ions:

- if the admixture and the main metal are discharged with a very low degree of polarisation, the precipitation of these components is determined by the rate of convective diffusion. If the valency of these metals is the same, the constant of convective diffusion is the same for both metals. The relative content of the admixtures in the cathode precipitate is directly proportional to its concentration in the solution;
- if the discharge of the main metal is accompanied by clearly visible polarisation, then a large amount of the admixtures is discharged at the threshold current density. This case is observed most frequently in practice. Since the rate of precipitation of the main metal is determined by the discharge stage, then for the same valency of the admixtures and the parent metal, the rate of precipitation of the admixture is higher and the result is a cathodic precipitate contaminated by the admixtures to a considerably greater extent than the electrolyte. The content of the admixtures in the cathodic precipitate during precipitation in the conditions of the threshold current density increases with a decrease of cathodic current density, and vice versa. Therefore, in the case of preliminary electrolysis at a low current density, it is possible to remove the majority of the admixtures without greatly depleting the electrolyte in the main metal, i.e. it is possible to purify the electrolyte to remove admixtures;
- in the case in which the rates of precipitation of the main metal and of the admixture are determined by the rate of discharge, the constant of the rate of discharge of the main metal must be higher than that of the admixture. However, if the constant of the rate

of discharge of the admixture is higher than that of the main metal, then the admixture, because of its low concentration in the solution, will be unavoidably discharged at the threshold current density and this may lead to contamination of the cathode precipitate with admixtures. If the constant of the rate of discharge of the ions of the main metal is higher than that of the admixture, the precipitate will be 'cleaner' than the electrolyte. In this case, the constant of discharge depends on the chemical individuality of the admixture;
- in another case, the limiting factor is diffusion of the main metal. This is the least favourable situation for producing pure metal. Therefore, it is necessary to carry out electrolysis at a very low current density in cases in which the admixture is less electropositive than the main metal.

In the electrolytic production of high-purity metals it is very important to carry out to deep purification of the electrolyte to remove admixtures of secondary cations. This is usually carried out by the electrochemical method of the so-called preliminary electrolysis of the electrolyte. In electrochemical purification of the electrolyte it is recommended to use the following methods:
- preliminary electrolysis at a low current density in which the threshold current results in the discharge of the cations of the more electropositive admixtures and their removal from the cathode region.
- electrolysis using an electrically charged diaphragm encircling the cathode. From the layer of the electrolyte in the vicinity of the cathode the cations of the more electropositive admixtures on the surface of the diaphragm are discharged, resulting in further precipitation of the pure metal on the cathode at a considerably higher current density. The diaphragm may be produced from, for example, perforated graphite.
- the purification of the electrolyte to remove the ions of the more electropositive admixtures may be carried out by cementation (i.e. internal electrolysis), which is the process of displacement of the ions of a more noble metal by another less noble metal from the solution, with the former being transferred from the ions to the metallic condition, and the latter vice versa. Cementation may be used for both producing noble metals from the solutions and for purification of electrolytes to remove admixtures. Suitable examples is the cementation of copper with iron or gold with zinc, displacement of copper and cadmium by zinc from a zinc electrolyte or of lead, tin, copper and nickel by indium from an indium electrolyte. Cementation takes place when the electrode potential of the displacing metal is far more electronegative than that of the metal being

displaced.

8.6. Removal of nonmetallic admixtures from cathodic metals in electrolysis

In order to obtain high-purity metals by electrolytic precipitation, it is necessary to form suitable conditions in which the cathode precipitate does not receive admixtures of metals, gases or metalloids. In electrolytic refining of metals of the group of iron, chromium, manganese and a number of other metals from aqueous solutions, the discharge of the cations of the metal is accompanied by the discharge of the hydrogen ions. The resultant hydrogen-charged metal must be subsequently processed by vacuum annealing or vacuum remelting. As regards the metalloids, S, Se, Te, C, Si and O can transfer to the cathode. Sulphur is present in the cathode metal in the form of sulphates or sulphides. The formation of the cathode layer of metals in the absence of S is possible only when using pure chloride or nitrate electrolytes. Carbon, silicon, selenides, tellurides and basic salts of arsenic, antimony, bismuth and also noble metals are transferred to the cathode only by means of the cataphoresis transfer of solid particles from the electrolyte on the surface of the cathode. The basic condition for producing high-purity metals in electrolysis is the complete optical transparency of the solutions. In justified cases, the electrolysis of pure metals is carried out using diaphragms, separating the cathode space from the anode space. The material of the diaphragms is represented by low-porosity cloth produced from plastics insoluble in the electrolyte or semi-transparent layers of colloidal cellophane.

It is very difficult to obtain an oxygen-free electrolytic precipitate. Oxygen is transferred into cathode layers in the form of hydroxides or in most cases in the form of oxides inclusions from the solution. The oxygen content of electrolytic copper varies from ten thousandths to thousandths of a percent. The removal of traces of oxygen can be carried out by the electrolysis of salt melts or by reduction remelting of the cathode precipitate in vacuum. Electrolytic precipitation of metals from salt melts at temperatures higher than the melting point of these melts creates more suitable conditions for producing metals of the highest purity.

8.7. Anodic dissolution. Electrolytic refining of metals

In anodic dissolution of an alloy consisting of several components, the metal with the highest electronegativity is the first to dissolve and is followed gradually by metals with lower electronegativity, etc, in accordance with the electrochemical voltage series (Table 10). Of course, this sequence may change depending on the content (activity) of each component in the alloy, and also in cases in which the alloy form eutectics, solid solutions or intermetallic compounds. In anodic dissolution of the alloy, the ionisation of the admixtures is more difficult. This is favourable for the production of high-purity metals by electrolytic refining of the initial metals containing a relatively small amount of metallic admixtures.

In electrolytic refining, the main metal is subjected to anodic dissolution with the formation of appropriate cations in the electrolyte which are discharged (precipitate) on the cathode in the metallic form in accordance with the reactions (72) and (74).

In the initial anodic alloy which is to be subjected to electrolytic refining, the main metal is usually characterised by higher electronegativity than the admixtures and, consequently, the admixtures do not anodically dissolved and buildup in the anode space. If the anodic metal contains admixtures with higher electronegativity than the parent metal, the admixtures will undergo preferential anodic dissolution and will be transformed into the electrolyte in the form of appropriate cations, provided solid solutions or intermetallic phases of these admixtures do not form. However, these admixtures do not discharge on the cathode in the presence of the ions of the main metal in the electrolyte, due to the fact that their electrode potential is more negative than the potential of the main metal.

Depending on the nature of metal, electrolytic refining may also be carried out in aqueous electrolytes and salt melts, using solid or liquid and also molten metals as electrodes. In aqueous electrolytes with solid electrodes, electrolytic refining is carried out on metals characterised by more electropositive electrode potentials in comparison with hydrogen (Cu, Ag, Ni, Co, Mn, Cr ...), and the soluble electrodes are represented by cast anode plates produced from commercial purity metal (Fig. 54).

Of considerable importance is the electrolytic refining of metals in salt melts. The absence of hydrogen ions in salt melts enables electrolytic refining of metals such as, for example, aluminium and magnesium. Aluminium is refined in the electrolytes formed by fluoride salts $NaF + AlF_3$, $Na_3AlF_6 + LiF$, etc, fluoride salts and also chloride salts $BaCl_2 + NaCl$. For the production of high-purity aluminium it is usually recommended to use a three-layer electrolysis system in which the anode is represented by, for example, an alloy of aluminium with copper with the intensity ~0.2 kg dm^{-3} higher than the intensity of the electrolyte. Pure cathode metal (with the intensity 0.2–0.3 kg dm^{-3} lower) flows on the surface of the electrolyte (Fig. 54b). Process temperature is usually 800 °C, i.e. this temperature is higher than the melting point of aluminium. Current density is 4000–5000 A m^{-2}. This method may be used to produce aluminium with the purity of up to 5N (99.999%) aluminium. The metal prepared by this method usually represents the starting material for further refining, for example, by zone melting.

Another efficient method is the electrolytic refining of salt melts of refractory metals at temperatures lower than the melting point of these metals. This relates in particular to Be, Ti, Zr, Nb. On the cathode, the metal precipitates in the crystalline form as a cathode sponge mass.

8.8. Amalgamation method of separation and refining of metals

The amalgamation method is based on the electrolysis of aqueous solutions with a mercury cathode. From electrolysis with solid cathodes, it differs by a relatively high excess voltage of precipitation of hydrogen on the mercury cathode. The metals, precipitating on the cathode, dissolve in mercury, i.e. they form amalgams, and the dissolution of certain metals is accompanied by the generation of heat. Consequently, the discharge potentials of their ions on the mercury cathode are more electropositive than the normal potentials. The combination of this phenomenon with the high excess voltage required for discharging H^+ ions on mercury also enables precipitation of alkali metals and alkali earth metals on the mercury cathode by electrolysis of aqueous solutions.

The formation of amalgams of nickel, cobalt, iron and chromium is very difficult and is accompanied by heat absorption. Therefore, the potentials of their precipitation on the mercury cathode are far more negative than the normal potentials. Consequently, it is possible to separate from them also more electropositive elements. The following do not precipitate from the acid aqueous solutions on the mercury cathode: Al, Be, B, Ti, U, Nb, Ta, V, Zr, P, Si, alkali metals, alkali earths and lanthanides. However, they can be separated from the elements, precipitating on the mercury cathode. Of course, this is only one of the aspects of the amalgamation process. The anodic process, i.e. dissociation of the amalgam by anodic dissolution together with the cathodic process provides extensive possibilities for using the amalgamation process in the separation and refining of metals.

The increase of the metal concentration in the amalgam or the increase of temperature result in significant changes of the anode potential. The special feature are the changes of the potential during the dissociation of the amalgams of Cr, Ni, Co and Fe. At room temperature, these metals become passive in dissolution in mercury, and with the temperature increasing to 78–83 °C the metals show depassivation and, consequently, the potentials of the dissolved metals are displaced by tenths of a volt to the side of electronegative values.

In amalgamation electrolysis, the values of the potential depend on current density, the concentration of the metal in the amalgam and on temperature. During electrolysis, the potentials are only slightly displaced to more electropositive values. As soon as the metal is almost completely transferred from the amalgam into the solution, there is a significant change of the potential to the values required for the oxidation of mercury. However, the almost complete precipitation of the metals from the amalgam takes place only at low current densities of 50–100 A m^{-2} and at a high current density there is a jump-like change of the potential, provided the amalgam still contains tens of milligrams of metal. If the amalgam contains two metals, after the transition of the metal with the more electronegative potential to the solution, the potential 'jumps' up to the value required for the oxidation of the next element. With increase of the content of the metal with the more electropositive potential in the amalgam, the amount of the metal with the electronegative potential remaining in the amalgam increases. Electrolysis with the mercury electrode is used in the refining of a number of metals, such as Ga, Tl, In, etc.

The electrolytic refining of metals by the amalgamation method is usually carried out in an electrolyser with bipolar mercury electrodes, Fig. 55. The electrolyser is produced from a non-conducting material and is separated by a diaphragm into two sections A and B in such a manner that the mercury in section A is not in contact with the mercury in section B. The mercury in the sections A and B is used as a bipolar electrode: in section I, mercury in section A operates as a cathode, in section II as an anode; similarly, in section B mercury is a cathode and in section III an anode. The electrolyte of section I contains the soluble anode from the pre-purified metal, and the electrolyte of section III contains a metallic or graphite electrode (cathode) on which the refined metal settles. The electrolyser may contains several bipolar electrodes, so that it is possible to increase the efficiency of refining of metals.

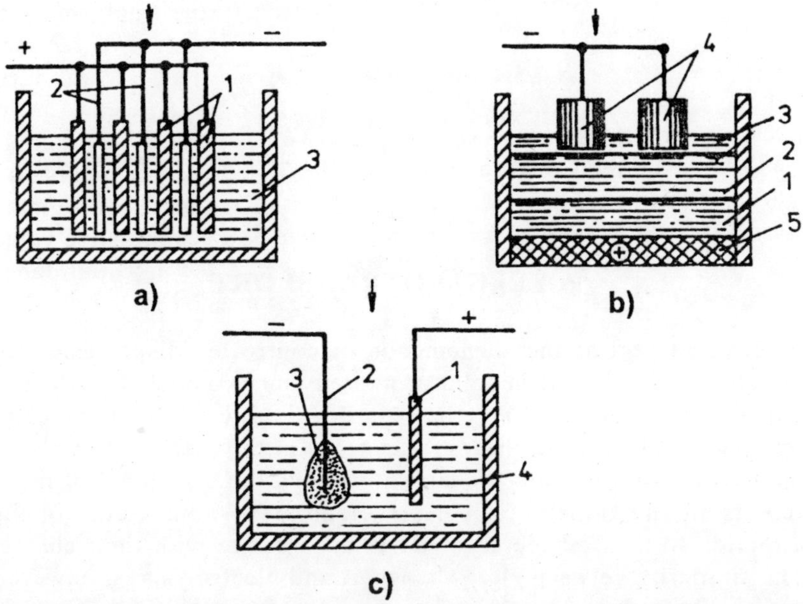

Fig. 54 Electrolysers for electrolytic refining of metals (electrolysis with soluble anodes). a) electrolysis in aqueous solutions with solid electrodes, 1) anodic alloy, 2) cathode (pure metal foil), 3) electrolyte (aqueous solution of metal salt); b) in salt melts with a liquid cathode and anode, 1) anodic alloy, 2) electrolyte (salt melt), 3) melt of pure cathode metal, 4) graphite cathodes, 5) conducting bottom; c) in salt melts with solid electrodes, 1) anode, 2) cathode, 3) refined metal, 4) electrolyte (salt melt).

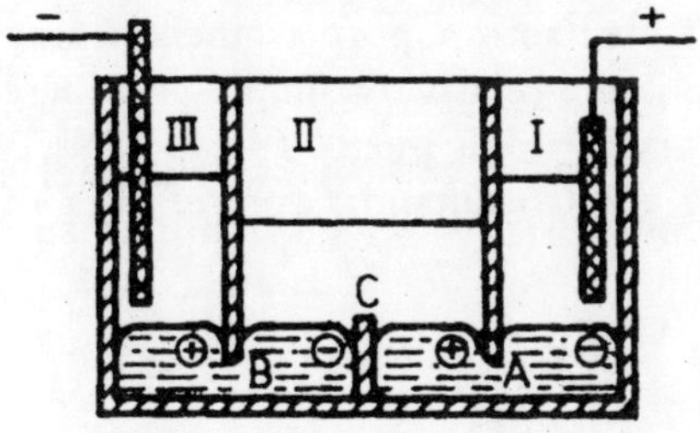

Fig. 55 Diagram of the electrolyser with bipolar mercury electrodes.

9. ELECTROTRANSPORT

Electrotransport as the phenomenon of controlled displacement of the atoms of elements in solid or molten metals under the effect of a direct electrical field may be regarded as one of the methods of refining metals. During electrotransport, as in the case of electrolysis, the passage of direct electric current through the specimens of metal results in the transfer of certain elements – admixtures in the direction to the cathode or anode in accordance with their charge. The similarity between electrotransport and electrolysis is, however, greatly limited, in particular, owing to the fact that the passage of electric current through the metallic specimen is determined by the directional motion of free electrons. Only a very small amount of electric current is transferred by ionised atoms of the main metal or admixtures present in them. The energy efficiency of electrotransport is consequently low and, therefore, to obtained a steep concentration gradient, it is necessary to ensure the long-term effect of direct current which is the main shortcoming of electrotransport as the refining method. However, a positive feature is the relatively simple procedure of the method which may result in very high purity (up to 8 N).

The main applications of using electrotransport for refining of metals are:

- electrotransport in solid metals
- electrotransport in metal melts
- electrotransport through a layer of oxides
- electrotransport in zone melting of metals.

Owing to the fact that a concentration gradient forms along the length of the specimen in refining of metal by electrotransport, the specimen is characterised by the occurrence, during the process, of reversed diffusion of the admixture component 1, and for diluted solutions it can be written that:

$$\frac{v_1}{D_1} = \frac{z_1 e}{kT} \quad (79)$$

where v_1 is the velocity, z_1 is the valency of the admixture component 1, e is the electron charge, k is the Boltzmann constant, D_1 is the diffusion coefficient, T is temperature.

However, in refining, the speed of motion of the admixture element 1 is not as important as the concentration exchange which the electric field E is capable of realising. This exchange is described quantitatively by expressing the flow of the substance under the effect of the field E. The equation has the following form

$$q_D = D_1 \frac{dC_1}{dl} + v_1 C_1 E \tag{80}$$

where q_D is the flow of the substance [mol cm^{-2} s^{-1}], C_1 is the is concentration of the admixture component 1 [mol cm^{-3}], l is the length to which the concentration C_1 corresponds, E is the intensity of the electric field [V cm^{-1}].

In the case of a sufficiently long time of the effect of the direct electrical field, the first member in equation (80) characterises diffusion and the second member the amount of the transported admixture component 1. Both these terms are balanced and provide the zero flow, corresponding to the maximum obtainable purity of the compound.

9.1. Refining of admixtures from solid metals by electrotransport

A metallic bar with length l_0 carries direct electric current which results in heating of the bar to the required temperature and, at the same time, in the transfer of the admixture, as indicated by Fig. 56. If τ_0 is the duration of the initial stage, corresponding to the initial concentration of admixture C_0, and the duration of electrotransport changes $\tau_1 < \tau_2 < \tau_\infty$ refining takes place in the initial part of the bar (Fig. 56). The maximum purity corresponds to the duration of

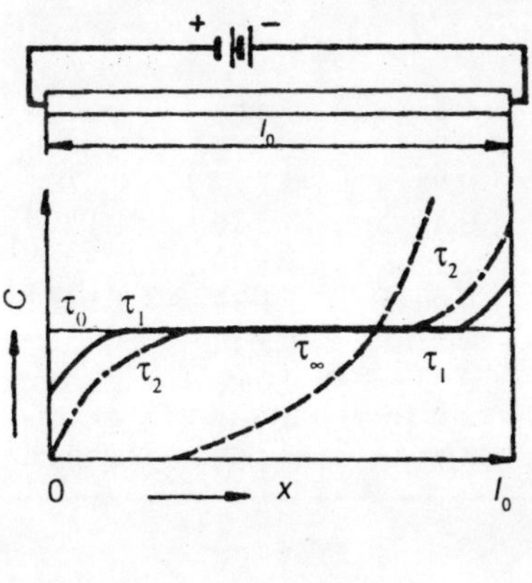

Fig. 56 Distribution of admixtures along the length of a metallic bar subjected to electrotransport in relation to the duration of the effect of direct current.

electrotransport required for the establishment of equilibrium, see equation (80).

The concentration of the admixture in the initial half of the length of the specimen generation to the initial concentration is defined by the general equation:

$$\frac{C_{m(\tau)}}{C_0} = \frac{2}{C_0 l_0} \int_0^{1/2} C(x,\tau)dx \tag{81}$$

where $C_{m(\tau)}$ is the concentration of the admixture in the initial half of the specimen, reached in the period τ, $C_{(x,\tau)}$ is the concentration of the admixture as the function of distance (x) from the beginning of the specimen and the duration of electrotransport (τ).

Equation (81) was used for processing the data on electrotransport of interstitial admixtures from the viewpoint of the possible refining of a number of metals, in particular, refractory metals to remove selected interstitial admixtures. In all cases, the length of the bar l_0 = 10 cm and the intensity of the electric field E = 0.2 V cm^{-1} were selected. The values of C_m/C_0 were calculated for a distance of 5 cm from the beginning of the bar for the duration after one day, after five days, and after the infinitely long time of electrotransport. The experimental results are presented in Table 11.

As indicated by Table 11, in all cases, with the exception of the Ti–O, Ta–C, W–C systems, there was deep refining already after electrotransport for 5 days. The identical method was used for the analysis of other systems for which reliable data are available in the literature. The calculations for the Co–C, Ti–C and Y–O systems show that deep refining takes place after 5 days of electrotransport, but in the case of the Fe–H, Fe–D, Ni–O and Ta–H systems, only the

Table 11 Relative concentration of interstitial admixtures in electrotransport in solid metals

System	Temperature [°C]	v_1/D_1 [cm^{-1}]	C_m/C_0 in the first half of bar		
			1 day	5 days	∞
Ti–O	1327	3.2	0.95	0.75	0.078
Pd–H	182	19	0.49	$3.5 \cdot 10^{-4}$	$1.1 \cdot 10^{-8}$
Th–N	1520	57	$2.1 \cdot 10^{-11}$	$3.3 \cdot 10^{-25}$	$3.3 \cdot 10^{-25}$
Th–O	1675	19	0.0045	$1.1 \cdot 10^{-3}$	$1.1 \cdot 10^{-8}$
Fe–C	1400	44	$4.2 \cdot 10^{-4}$	$1.5 \cdot 10^{-19}$	$1.5 \cdot 10^{-19}$
Ni–C	1400	12	0.73	0.065	$1.2 \cdot 10^{-5}$
Ta–C	2600	52	$4.9 \cdot 10^{-23}$	$4.9 \cdot 10^{-23}$	$4.9 \cdot 10^{-23}$
W–C	2600	5.7	0.75	0.14	0.0067

medium degree of refining is obtained.

The results of a number of experiments show that the v_1/D_1 ratio for the given system decreases with increasing temperature. Since the result of refining is proportional to the value of v_1/D_1, the theoretically attainable degree of refining should increase with decreasing temperature. However, on the other hand, the mobility of ionised atoms in solid metals increases with increasing temperature. Consequently, in experiments it is always necessary to define the temperature resulting in optimum refining conditions. Since the electrotransport in solid metals takes place in defined conditions, it is possible to obtain, by electrotransport, the efficient removal of interstitial admixtures in certain metals. In the case of substitutional elements, it is possible to obtain a high degree of refining only after a very long process time. This restricts the range of application of electrotransport in practice. However, electrotransport in solid substances is also regarded as a long-term process taking place in practice in the presence of the direct field. Consequently, electrotransport contributes to, for example, 'ageing' of semiconductor components.

9.2. Refining of admixtures from metallic melts by electrotransport

The electrotransport of metal melts is carried out using equipment consisting of a pipe sealed at one end and filled with metal melts, whereas the other end of the pipe is connected with a container holding the metallic melt. After applying a low-intensity direct field diffusion transfer of certain admixtures takes place in the melts. The concentration of the admixtures in the section of the metal situated in the pipe which is in contact with the container, remains, however, almost completely constant, if the container is sufficiently large and the thermal convection takes place.

The electrotransport of admixture elements in selected melts of low-melting metals was investigated by Belashchenko. He obtained the values of v_1/D_1, presented in Table 12.

All the dissolved metals are characterised by the value $v_1/D_1 > 12\ cm^{-1}$ and, consequently, they can be easily removed after 5 days of electrotransport. Nevertheless, it is important to consider the process of thermal convection, decreasing the efficiency of purification. The efficiency of convection increases with increasing intensity of the electric field.

The electrotransport in metallic melts at the intensity of the electric field of 0.05 V cm^{-1} or lower (in order to reduce the intensity of

Table 12 Values of v_1/D_1 in electrotransport of admixtures in melts of low-melting metals, t = 300–350°C, E = 0.02 V cm^{-1}

Solvent	Dissolved metal	v_1/D_1 [cm^{-1}]
Cd	Ni, Co, Te, Ag, Au, Bi, Pb, Sn, In	23
In	Ni, Co, Te, Ag	14
Sn	Ni, Co, Te, Ag, Au, Sb, Cd, Ca	14
Pb	Co, Te, Ag. Au, Sb, Cd, Se, Sn	1.9
Bi	Ni, Co, Te, Ag, Au, Cd, Se, Sn	3.7

convection) and the appropriate length of the pipe (> 20 cm) may be utilised sufficiently for removing a number of dissolved admixtures from metallic melts. In particular, the admixtures with high values of v_1/D_1 are removed easily. However, electrotransport is characterised by low efficiency in the separation of the isotopes of the same metal in metallic melts.

9.3. Refining of metals by electrotransport through the oxide layer

This type of electrotransport may be compared to a certain degree with the electrolytic refining of metals in aqueous solutions because, in this case, the electrolyte is represented by the solid layer of the metal oxide. A number of experiments has been carried out with electrotransport through the layer of iron oxides. Iron specimens were subjected to partial oxidation in order to produce a layer of FeO with a thickness of approximately 0.6 mm on two flat surfaces. The specimen was placed between two iron sheets (Fig. 57), heated to 1000 °C, followed by passing direct current with a density of 50 A cm^{-2} for 24 hours. The cross section of the specimen after the passage of current was used to examine the layer of deposited iron on longitudinal planes AB and A'B', and the thickness of the layer was proportional to the duration of the experiment and the density of current passing through the specimen. On the other hand, thickness decreased as a result of decay of anodic sheets of the planes CD and C'D'. The passage of direct current along the planes AB and A'B' resulted in the formation of a cathodic layer and in the dissolution of the anodic sheets along the planes CD and C'D'.

The mechanism of electrotransport through the layer of FeO may be described probably as follows: the initial stage is characterised by the migration of the Fe^{2+} ion from the anodic surface under the effect of the electric field through the lattice of FeO to the cathodic

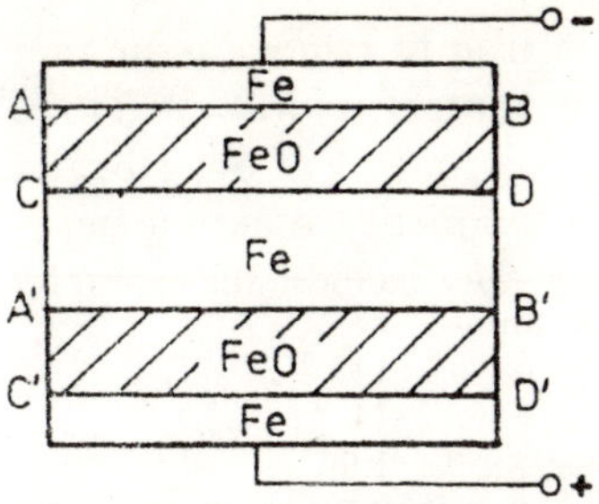

Fig. 57 Experimental setup in electrotransport through the iron oxide layer.

surface where it receives two negative charges and changes to the iron atom; correspondingly, the given area of the lattice of FeO is characterised by the formation of a hole which, in contrast to the Fe^{2+} ion, moves in the direction of the anode surface. The iron atom of the anode plate enters into reaction with the hole and releases two negative electrical charges. The anode plate dissolved and the resulting Fe^{2+} ion travels to the cathode surface, etc.

Refining of iron to remove admixtures by means of electrotransport through the FeO layer was carried out using electrolytic iron with a purity of 99.988% Fe as anodic metal. The experiments at a current density of 70 A cm^{-2}, a temperature of 985 °C, duration 70 hours, resulted in a cathodic deposit with a thickness of 2 mm, analysed in a mass spectrograph. Table 13 shows the results of analysis in comparison with the analysis of initial anodic iron.

Table 13 shows that by electrotransport through the layer of FeO

Table 13 Analysis of initial iron and iron after refining by electrotransport [at. ppm]

Admixture	Initial metal	Refined metal	Admixture	Initial metal	Refined metal
Ta	2	2	V	2	0.1
Sb	0.4	0.2	Ti	30	3
Pb	2	0.7	Ca	10	10
Ag	0.4	0.4	K	1	1
Mo	3	0.3	Cl	200	200
Nb	0.3	0.1	S	400	40
Zr	0.4	0.4	P	70	1
Sc	4	4	Si	750	750
As	4	1	Mg	10	10
Ga	4	1	Na	0.3	0.3
Ni	50	50	B	1	1
Co	70	70	O	350	–
Mn	30	0.7	C	170	–
Cr	80	10			

it is possible to purify efficiently metallic iron to remove a number of admixture elements. For example, the content of Mo, V, Ti, S, P decreases by more than a factor of 10, the content of Sn, Nb, As, Ga, Cr, Sb by a factor of 2–3. However, a number of other elements remain unchanged, even after electrotransport; they include, for example, Ni, Si, Co, Na, B, Ca.

There are two factors which may be used to explain the efficiency of purification by electrotransport through the layer of FeO:

- the insolubility of certain elements in the solid layer of the iron oxides resulting in the buildup of these elements at the anode-oxide interface (S, P, As);
- the admixtures dissolving in FeO are characterised by a lower speed of motion of the ions in the FeO lattice in comparison with the speed of motion of the Fe^{2+} ion.

The identical process of electrotransport through the layer of oxides may also be used for other metals whose oxides are characterised by electrical conductivity at the temperature lower than the melting point of the metal. The advantage of this purification method is the fact that there is no danger of contamination of the refined metal by the material of the boat, the atmosphere, etc., because refining is carried out in the solid state. However, the metal to be purified always contains a certain amount of oxides which may be removed only subsequently, for example, by vacuum remelting of metals. However, at the same time, it must be remembered that similar processes of the redistribution of the admixtures by electrotransport in the solid state through the oxide layer may also take place in technical practice as undesirable processes, sometimes during accidental occurrence of a direct current field. Of course, this has a negative effect in changing the homogeneity of the substance and increases, for example, the susceptibility to corrosion.

9.4. Applications of electrotransport in zone melting

The passage of direct current through a metallic bar refined by zone melting results in most cases in the positive effect of refining, despite the fact that the proportion of the current resulting in the displacement of ionised atoms in the metal is not large. Since zone melting is a long-term process, the duration of the process is to a certain extent comparable with the duration of the process of electrotransport and, consequently, the application of electrotransport is advantageous [3].

As regards the investigations carried out in this area, it is important to mention the method of electron beam zone melting of tungsten

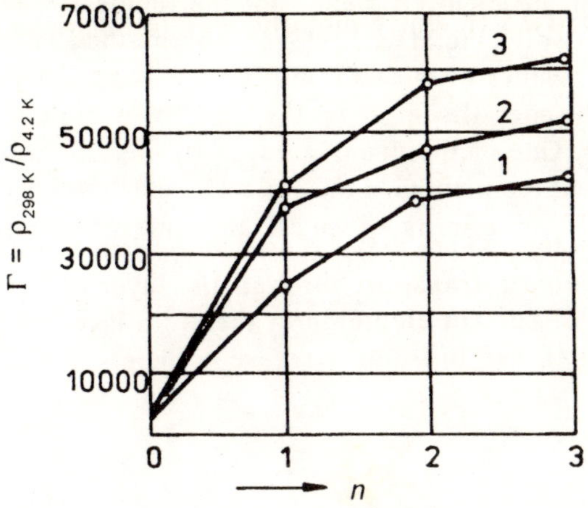

Fig. 58 Refining of W by electron beam zone melting combined with electrotransport (1 – E = 0 V cm^{-1}; 2 – E = 0.125 V cm^{-1}; 3 – E = 0.25 V cm^{-1}).

than with the application of the electric field of direct current using the specimens of tungsten wire with a diameter of 1.5 mm, 350 mm long. In the experiments, the number of passages of the molten zone and the intensity and direction of the electric field were varied. The results of experiments were evaluated by measuring the ratio of the resistance of tungsten wire specimens at room temperature and at the liquid helium temperature $\rho_{298\,K}/\rho_{4.2\,K}$, Fig. 58.

Figure 58 shows that the additional direct electric field improves in all cases the efficiency of refining of tungsten and the ratio of the resistance values increases. A positive effect is also exerted by the increase of the intensity of the electric field from 0.125 to 0.250 V cm^{-1}, corresponding to a current density of 500 and 1000 A cm^{-2}. However, the positive effect of the additional field is evident only when the field is oriented in the direction against the direction of movement of the zone. The field oriented in the direction of the movement of the zone has a negative effect on the purity of tungsten.

Experiments were carried out with the refining of cerium by electrotransport combined with electron beam zone melting. In simple electrotransport at 700–750 °C, the admixtures Fe, Cu, Si and O are displaced to the anode. In the zone melting of cerium without electrotransport, the admixtures Cu, Si and Fe are transferred in accordance with their values of k <1 to the end of the ingot.

It may be concluded that electrotransport as the refining method may be used with sufficient efficiency for the removal of only certain admixtures from solid metals. However, the method is efficient in refining a large number of metallic melts, if it is possible to prevent convection in the metallic melt in the case of a high intensity of the electric field. One of the disadvantages of electrotransport is the low productivity of the method. Consequently, electrotransport is used preferentially in the refining of small amounts of metals where a high degree of purity is required.

10. DIFFUSION

10. 1. Separation of gas substances by diffusion and thermal diffusion

The separation and purification of gas components may be carried out utilising their different rates of diffusion through porous barriers (membranes). Gases with higher diffusion rates concentrate on one side of the membrane, on the other side gases with lower diffusion rates. Multiple repetition of this process in the chamber with a large number of porous membranes makes it possible to obtain a high degree of separation of the gases.

At a low concentration of the gases diffusing through the porous membranes (when the dimensions of the pores are small in comparison with the free path of the molecules) the ratio of the rates of diffusion v_A and v_B of any gases A and B at the given temperature and pressure is inversely proportional to the ratio of the density ρ of these gases or their molecular masses M_A, M_B

$$\frac{v_A}{v_B} = \sqrt{\frac{\rho_B}{\rho_A}} = \sqrt{\frac{M_B}{M_A}} \qquad (82)$$

Therefore, as the gas becomes heavier or the molecular mass of the gas increases, the rate of diffusion of the gas through the porous membrane decreases and, consequently, the concentration of the gas behind the membrane decreases.

Another form of utilisation of the diffusion processes for separation of the gas components is thermal diffusion. If a container with a mixture of gases is heated on one side and on the other side it is cooled, there will be a difference between the composition of the mixture in the hot and cold zones. The differences in the composition increase with an increase of the temperature gradient between the two zones. The hot zone is characterised by the concentration of the components with lower molecular mass and vice versa.

Diffusion through porous membranes (diffusion cascades) and thermal diffusion are used in practice in particular in the separation of isotopes in the form of gas compounds, for example ${}^{235}_{92}UF_6$ from ${}^{238}_{92}UF_6$ in the enrichment process (despite the fact that ultracentrifuges are used in this case).

10.2. High-voltage electrodialysis

High-voltage electrodialysis is the process of deep refining of substances, for example, electrolytes, from which admixtures are precipitated by the application of an external direct high-voltage electrical field [3]. Electrodialysis is based on the acceleration of diffusion of ions through a semi-permeable membrane. The direct electrical field makes it possible to increase greatly the degree and rate of purification of substances.

Figure 59 shows the schematic section through electrodialysis equipment consisting of a vessel separated by membranes or diaphragms into three chambers. The membranes are permeable to the electrolyte and impermeable to the solid refined phase. The central chamber contains a suspension of the refined substance in water, the outer chambers contain pure water and electrodes. After application of the direct external electric potential to the individual electrodes, the anions from the central chamber penetrate through the membrane to the positively charged electrode and the cations penetrate the negatively charged electrode. Basis compounds form in the cathode chamber, acids build up in the anode chamber. According to the degree of saturation in the outer chambers, the solutions are discharged from them and the chambers are filled with pure ionised water. This makes

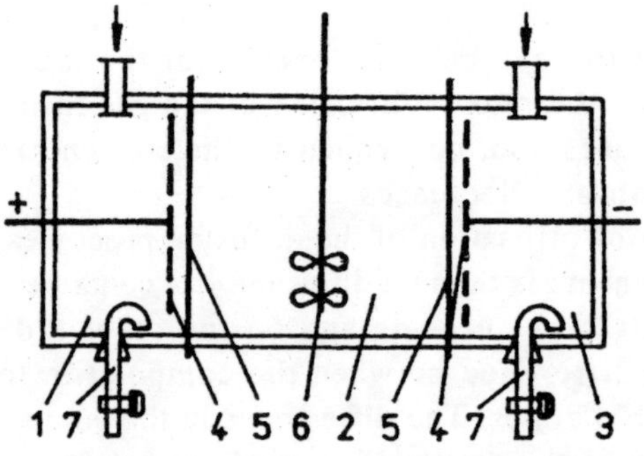

Fig. 59 Scheme of a three-chamber electrodialyser (1 – anode chamber, 2 – middle chamber, 3 – cathode chamber, 4 – platinum electrode, 5 – diaphragm, 6 – stirrer, 7 – seal).

it possible to increase the rate of removal of the admixtures from the electrolyte in the central chamber.

The process of electrodialysis is characterised by a relatively high rate and efficiency at a potential gradient of 150–250V cm^{-1} and at the total voltage on the terminals of equipment of 1500–1800 V. Current usually does not exceeds units of amperes. The efficiency of the process depends not only on the magnitude of high voltage but also on the rate of exchange of water in the chambers. At high concentrations of the ions in the anode or cathode chamber, reverse diffusion of ions into the central chamber may take place. In the case of 'weak' electrolytes, the efficiency also depends on the degree of dissociation of the electrolyte. The efficiency of high-voltage dialysis may be increased by, for example, the application of a five-chamber electrodialyser where two other small chambers are used as collectors of precipitated and removed ions of the admixtures thus preventing reversed diffusion.

High-voltage electrodialysis is used in the refining of a number of substances such as, for example, SiO_2, TiO_2, $NiCO_3$, MgO, to remove the admixtures present in the electrolyte, such as sulphides, chlorides NaCl, KCl, $CaCl_2$, $MgCl_2$. The removal of the admixtures insoluble in water, for example, iron, aluminium, copper, is carried out using complexing substances which selectively transfer the admixtures to the soluble condition. In biological applications, dialysis units include, for example, artificial kidney components.

11. SELECTIVE PRECIPITATION, OXIDATION AND REDUCTION

In selective precipitation from solutions of low-solubility compounds, the sequence of precipitation of these compounds is determined by changes of the isobaric potential of the process. In the first instance, precipitation takes place in metals in which the formation of low-solubility compounds is accompanied by the maximum decrease of free energy or isobaric potential ΔZ^0.

It is assumed that the components A and B that are in a solution form a low-solubility compound A_mB_n in accordance with the reaction

$$mA + nB = A_mB_n \tag{83}$$

The equilibrium of the reaction at a given temperature is established on the condition that the sum of the potentials of the initial compounds is equal to the isobaric potential of the final product. The isobaric potential of the solid phase with a constant composition is consequently constant. The reaction of precipitation of low-solubility compounds A_mB_n takes place if the activity (concentration) of the initial compounds A and B in the solution (or in the melt) satisfies the condition

$$m\Delta Z_A^0 + mRT \ln a_A + n\Delta Z_B^0 + nRT \ln a_B = \Delta Z_{A_mB_n}^0 \tag{84}$$

where ΔZ^0 is the change of the standard isobaric potential, a_A, a_B is the activity of the components, and the activity of the solid phase (compound) A_mB_n is assumed to be equal to 1. The relationship between the activities of the initial components in the solution in equilibrium with the precipitated low-solubility compound is expressed by the equation

$$\ln a_A = \frac{\left(\Delta Z_{A_mB_n}^0 - m\Delta Z_A^0 - n\Delta Z_B^0\right)}{nRT} - n \ln \frac{a_B}{m} \tag{85}$$

The equation also shows

$$a_A^m a_B^n = \exp\left(\frac{\Delta Z_{A_mB_n}^0 - m\Delta Z_B^0 - n\Delta Z_B^0}{RT}\right) = \exp\left(\frac{\Delta Z_p^0}{RT}\right) \tag{86}$$

where a_A^m, a_B^m is the susceptibility of the given compound to solubility, i.e. the stability of the given compound, ΔZ_p^0 is the change of the isobaric potential of the reaction in the standard conditions.

It is evident that as the absolute value of the standard isobaric potential of the reaction of formation of the low-solubility compound increases, the solubility of the resultant compound decreases.

A suitable example is the reaction of formation of hydroxides because for the majority of metals they are low-solubility compounds, Table 14:

$$Me^{n+} + n\ OH^- \rightleftarrows Me(OH)_{n[S]} \tag{87}$$

In precipitation of hydroxides of metals from the solutions, the hydroxides of the metal with the lowest pH (more acid solution) and also those with lower solubility are the first to precipitate. This relationship, determining the sequence of precipitation of salts, is also valid for other compounds. In precipitation of two salts from the solution with the same anion, the ratio of the content of two compounds in the solid phase characterised by isomorphous precipitation is proportional to the ratio of the concentrations of these compounds in the solution.

Selective reduction or oxidation, i.e. transition of ions to lower or higher valency, may be used to increase the differences in the properties of similar elements, for example, facilitate the transition of one of the elements into the solution or decrease its volatility

Table 14 Equilibrium pH values of formation of hydroxides of several metals

Hydrolysis reaction	ΔZ_p^0 [kJ g_{eq}]	$a_A^m a_B^n$	Solubility [mol l^{-1}]	pH
$Co^{3+} + 3H_2O = Co(OH)_3 + 3H^+$	−77.5	$3.0 \cdot 10^{-41}$	$5.7 \cdot 10^{-11}$	1.0
$Sb^{3+} + 3H_2O = Sb(OH)_3 + 3H^+$	−73.2	$4.0 \cdot 10^{-42}$	$1.1 \cdot 10^{-11}$	1.2
$Sn^{2+} + 2H_2O = Sn(OH)_2 + 2H^+$	−72.0	$5.0 \cdot 10^{-26}$	$2.3 \cdot 10^{-9}$	1.4
$Fe^{3+} + 3H_2O = Fe(OH)_3 + 3H^+$	−71.2	$4.0 \cdot 10^{-38}$	$2.0 \cdot 10^{-10}$	1.6
$Al^{3+} + 3H_2O = Al(OH)_3 + 3H^+$	−62.4	$1.9 \cdot 10^{-33}$	$2.9 \cdot 10^{-9}$	3.1
$Bi^{3+} + 3H_2O = Bi(OH)_3 + 3H^+$	−57.8	$6.3 \cdot 10^{-33}$	$6.3 \cdot 10^{-9}$	3.9
$Cu^{2+} + 2H_2O = Cu(OH)_2 + 2H^+$	−57.8	$5.6 \cdot 10^{-20}$	$2.4 \cdot 10^{-7}$	4.5
$Zn^{2+} + 2H_2O = Zn(OH)_2 + 2H^+$	−46.5	$4.5 \cdot 10^{-17}$	$2.2 \cdot 10^{-6}$	5.9
$Ni^{2+} + 2H_2O = Ni(OH)_2 + 2H^+$	−39.4	$1.0 \cdot 10^{-15}$	$1.4 \cdot 10^{-5}$	7.1
$Mg^{2+} + 2H_2O = Mg(OH)_2 + 2H^+$	−32.2	$5.5 \cdot 10^{-12}$	$1.1 \cdot 10^{-4}$	8.4

ΔZ_p^0 – change of the standard isobaric potential of the hydrolysis reaction
pH – hydrogen factor of formation of hydrates which depends on individual properties and on activity of ions of precipitated metal in solution
$a_A^m a_B^n$ – stability of the compound

in comparison with the other element. For example, in the extraction of germanium tetrachloride $GeCl_4$ by hydrochloric acid, it is even possible to extract all admixtures, with the exception of $AsCl_3$. However, the solubility of arsenic rapidly increases if the As^{3+} ion is transferred by oxidation to As^{5+}. This may be carried out by adding chlorides or nitric acid. The resultant $AsCl_5$ is almost completely soluble in concentrated hydrochloric acid.

The separation of zirconium from hafnium is carried out utilising differences in the volatility of $ZrCl_4$ and $ZrCl_3$. In heating of $ZrCl_4$, containing admixture of $HfCl_4$ with zirconium powder in vacuum to 400–450 °C, the majority of zirconium tetrachloride is reduced to $ZrCl_3$ with relatively low volatility. $HfCl_4$ is not reduced and at the same time may be obtained in the form of a hafnium concentrate with the ratio Hf : Zr > 10 together with part of the non-reduced $ZrCl_4$. In the remaining $ZrCl_3$, the Hf : Zr ratio decreases from 2 to 0.05. Subsequently, $ZrCl_3$ can be transferred back to $ZrCl_4$ at a temperature of 550 °C. $ZrCl_4$ contains less hafnium than the initial specimen.

Selective oxidation and reduction of metals, of their oxides or other compounds in classic pyrometallurgical reduction and refining processes [43], including deoxidation processes in the refining melting of metals and alloys, are well known from the theory of metallurgical processes.

12. DEGASIFICATION OF METALS

In the production of pure metals, it is necessary to pay special attention not only to the presence of gases in them but also to the methods and means of decreasing the concentration of gases dissolved in metals. The presence of the gases nitrogen, oxygen, hydrogen and also of carbon and other interstitial admixtures in metals, even in very small quantities, has a strong negative effect on the physical and mechanical properties of metals, especially plasticity and brittleness, for example U, Ti, Zr, V, Ta, Cr, W and Mo. Therefore, sections of electrical vacuum devices operating in a high vacuum of 10^{-3} up to 10^{-6} Pa and at higher temperatures are made of structural material such as, for example, Cu, W, Mo, Ta where the maximum permissible content of gases is limited.

12.1. Gases in metals

The nature of gases present in metals, i.e. the nature and strength of bonding of gases with metals, may differ:

- adsorption of gases by the surface of metals (physical adsorption and chemisorption);
- dissolution of gases in metals;
- chemical reaction of gases with metals.

The diffusion of gases is a process leading to the equilibrium condition by the equalisation of the gas pressure in the metal, both in the solid state and in melting in the given medium. The rate of diffusion of gases in metals for diatomic gases oxygen, hydrogen and nitrogen may be expressed by the equation

$$f_D = \frac{np}{d}\exp\left(-\frac{\Delta H_D}{2RT}\right) \tag{88}$$

where

- f_D is the amount of the gas diffusing at partial pressure p through the unit area of metal during unit time at a temperature T;
- ΔH_D is the change of enthalpy of diffusion in the metal–gas system;

- n is a constant;
- d is the thickness of the metal layer.

This equation indicates that the rate of diffusion of the gas is inversely proportional to the thickness of the metal layer through which the gas passes. The rate of diffusion increases with increasing temperature and gas pressure. The formation of oxides and other layers changes the dependence of diffusion on pressure. When a compact 'protective' layer forms on the surface of metal, diffusion is no longer dependent on pressure.

The diffusion of gases takes place preferentially and directly through the crystal lattice of the metal, not along the grain boundaries. The gases diffusing into the metals are capable of dissolving in them. They are implanted in the lattice with the formation of solid solutions, mostly interstitial. In the majority of cases, the dissolution of gases in metals is an endothermic process and, consequently, the solubility of gases in metals usually increases with increasing temperature. During cooling of metals, the gases escape from the metals and this results in a number of difficulties in both crystallisation (formation of bubbles, etc.) and in solid state transformations (flakes in steel, etc.).

The temperature dependence of the isobaric solubility of diatomic gases in metal is described by the equation

$$s = \text{const. } \exp\left(-\frac{\Delta H_S}{2RT}\right) \tag{89}$$

and the dependence of solubility on pressure (according to Sieverts law) is expressed by the equation:

$$s = \text{const.} \sqrt{p} \tag{90}$$

where ΔH_S is the molar enthalpy of dissolution of the gas; s is the solubility of the gas; p is the partial pressure of gas above the melt.

The square root of partial pressure p in equation (90) expresses the dissociation of the gas molecule into two atoms. The cube root is used in equation (90) for a molecule with three atoms, etc. The process of dissolution of gas may take place at positive or negative values of the enthalpy of dissolution ΔH_S, i.e. the process may be endothermic or exothermic. In the endothermic dissolution process, the solubility of gas in metal increases with increasing temperature, Fig. 60.

In the molten state, the majority of metals dissolve oxygen which forms oxides of, for example, copper, cobalt, silver, vanadium, titanium, etc. However, oxygen does not dissolve in noble metals, gold, mercury and platinum. Oxidation of metals may also take place as a result

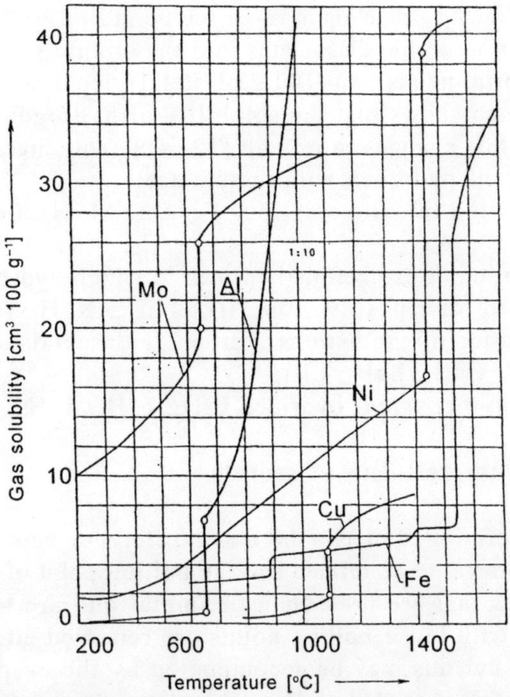

Fig. 60 Dependence of the solubility of hydrogen in some metals on temperature (endothermic dissolution process).

of reactions with the vapours of water or other gases CO, CO_2, SO_2. More complicated gases, containing oxygen, dissociate into atoms prior to dissolution in the metal. During cooling of the metal, the excess oxygen precipitates as an oxide or free gas, for example, in silver.

Nitrogen is characterised by preferential dissolution in metals already at low temperatures. At higher temperatures, nitrogen forms nitrides with the metallic elements Fe, Mn, Al, Mg, Cr, V, W, Mo, Zr, Ti ... At temperatures below 1200 °C, nitrogen is almost insoluble in copper, nickel and cobalt. Aluminium is capable of dissolving at 700 °C 1 cm^3 of nitrogen in 100 g of metal which represents $\sim 1.2 \cdot 10^{-3}$ wt.% of nitrogen in aluminium. The nitrogen, formed in the metals, can be divided into three groups:

– ion (Li_3N, K_3N, Be_3N_2, Ca_3N_2)
– covalent (AlN, BN, Si_3N_4)
– admixture (TiN).

Hydrogen dissolves in all technically important metals and is the only gas diffusing in metals at room temperature. Prior to dissolution, the hydrogen molecule dissociates into the atomic gas. The solubility of hydrogen in metals ($cm^3/100$ g of metal) depends on temperature, Fig. 60, which shows that the solubility of hydrogen at the melting point of metals changes in a jump [13, 41]. The characteristic types of reactions of hydrogen with metals are:

- the formation of solid solutions Fe, Cu, Al, Ni, Co, Cr, Mo, Ag, Pt;
- the formation of exothermic hydrides in which hydrogen precipitate in the form of a positive ion TiH_2, LaH_3, NbH, TaH, PdH, VH;
- the formation of exothermic hydrides with metals of groups I and II – NaH, CaH_2, BaH_2, LiH;
- endothermic covalent hydrides B_2H_5, SiH_4, TeH_2.

12.2. Vacuum refining of metals

In vacuum refining of metals there is a difference between the boiling point of the metal to be refined and the boiling point of the admixtures or impurities that are present in the metal and are to be removed. The metals with lower boiling points are removed either by vacuum distillation, but this may be accompanied by the evaporation of not only the metals regarded as impurities but also metals regarded as alloying elements, for example, chromium in electron beam or plasma melting of W–Cr–Co alloys or manganese and copper in vacuum melting of other high-temperature alloys. Consequently, these alloys cannot be efficiently and reproducibly melted without large losses of evaporating or sublimating metals or compounds.

In practice, there are sometimes cases in which a large reduction of pressure results directly in the sublimation of refined metal. In non-ferrous metallurgy, this method is mainly used for refining of, for example, calcium, high-purity magnesium for the reduction of uranium, or tellurium for application in semiconductor compounds, etc.

In vacuum, the metals are refined not only by distillation in the elemental condition but also by means of easily evaporating or dissociating compounds.

With a decrease of the total pressure in the gas phase, the equilibrium and actual amount of dissolved gas in metal also decreases. In the presence of total vacuum, the gas content of the metal should be equal to zero. However, this applies only to a very thin layer of the melt. The gas may escape from the melt in vacuum only from the surface layer. Inside the melt, the high metallic static pressure prevents the release of gas in the form of bubbles. Therefore, in this case it applies that sufficiently effective and rapid degassing of the melt is possible only when the surface of the melt is renewed by suitable movement of the melt. Otherwise, the gas can travel to the surface only by means of diffusion and this greatly reduces the rate of the process.

Bubbling of the melt with inert gases is an efficient method of not only mixing the melt but also reducing the partial pressure of the gases dissolved in the melt which may diffuse from these bubbles into the melt during rising of the bubbles of the inert gas through the melt.

The gases forming compounds with a metal, for example, Cu_2O in copper, Al_2O_3 or $AlCl_3$ in aluminium, etc., can be removed only to a limited extent by the effect of vacuum. The amount of the gases remaining in the melt is determined by the equilibrium of their dissociation:

$$2\,\mathrm{MeX}_{[L]} = 2\,\mathrm{Me}_{[L]} + \mathrm{X}_{2[G]} \qquad (91)$$

where Me is some bivalent metal, and X is the diatomic gas.

Of the metals forming compounds with metals, the most important is oxygen whose removal from metal is referred to as deoxidation. However, deoxidation cannot be carried out exclusively in vacuum, and its efficiency may be increased by adding a suitable reduction agent, for example, carbon, whose product is the gaseous carbon oxide and vacuum is very important for its removal.

The most radical method of decreasing the total level of the gases in metals is vacuum zone melting at a very low residual pressure (lower than 1 Pa) in the melting chamber. Other refining methods, for example, distillation in nitrogen, melting in an induction furnace, electrolysis, etc., do not enable significant changes in the content of gases in the metals [43].

Electron beam vacuum melting is a highly efficient method of degassing metals enabling a very rapid decrease of the gas content in both non-ferrous metals and, in particular, in refractory metals W, Mo, Ta, Nb, Ti, Zr, Re and their alloys. Electron beam heating is utilised in particular in zone refining and melting of these alloys and metals. Electron beam melting is carried out at temperatures of up to 5000 °C and at a working vacuum of 10^{-2}–10^{-4} Pa. The process enables a large decrease of the gas content and removal of easily volatile compounds from the main electron beam remelted metal. In electron beam melting of the metal to be used for refining, the droplets of molten metal fall into a water-cooled solidification mould. In the working vacuum, there is not only removal of gases oxygen, hydrogen, and nitrogen from the metal, but the high working temperature also results in the operation of metal admixtures with high vapour tension (Mg, Zn, Pb, Cu, Sn). In addition to this, the carbon content decreases as a result of its reactions with the residual oxygen present in the metal with the formation of gaseous carbon monoxide. Further refining of the metal also takes place by the evaporation of volatile admixtures from the surface of molten metal in the solidification mould. Consequently, it is possible to remove 30-50% of volatile admixtures from the metal bath.

In refining of refractory metals by electron beam vacuum melting, the total content of the gases present in the melt decreases by approximately a factor of 100, the oxygen content decreases from 10^{-2} to 10^{-4}%, and the concentration of the admixtures nickel, iron and also of other metals decreases by approximately an order of magnitude. This refining process results in a large increase of the plasticity of metals and in a large (15–20 times) increase of the durability of structural parts of electronic and electric vacuum systems.

13. METHODS OF TESTING PURE METALS

13.1. Analytical methods

The methods of analytical chemistry occupy a significant position in the inspection of the purity and quality of metals, alloys and compounds. The main task in the analysis of pure substances is the qualitative and quantitative determination of the content of admixtures and accompanying elements with maximum accuracy. Consequently, the analytical methods are not only the means of inspection but also the integral part of every research or industrial practice, and also in other areas, in addition to chemical production.

The currently available methods of the chemical analysis of pure metals will be analysed below. The maximum efficiency of analysis can be achieved by the combined application of different chemical analytical methods of determining the content of elements; this is given in Table 15, including the sensitivity of these methods [47].

Table 15 Range of application of some analytical methods

Method	Application range [%]
Gravimetry, titration	100 to 10^{-2}
Polarography	1 to 10^{-5}
Emission spectroscopy	1 to 10^{-6}
Flame spectroscopy	1 to 10^{-8}
Absorption spectrophotometry	1 to 10^{-6}
Atomic absorption photometry	1 to 10^{-8}
Fluorimetry	1 to 10^{-8}
Mass spectrometry	10^{-3} to 10^{-10}
Activation analysis	10^{-3} to 10^{-12}

The analytical determination of the content of elements in specimens may be divided into the following stages:
- sampling, preparation and dissociation of specimens.
- separating by the chemical method or on the basis of different physical principles;
- the actual determination of the content of admixtures, measurement of the selected physical or chemical properties.

13.2. Gravimetry and titration

The principle of gravimetry (weight analysis): the component of the specimen to be investigated is transferred into a low-solubility reaction product and the amount of the product is determined by after conversion to the substance of the defined composition by drying or baking.

The separation of the component to be determined may be carried out by the following procedure:
- leaching with water or another solution agent
- expulsion of the component to be determined by increasing temperature (annealing, calcination).
- electrolytic precipitation of metals on a platinum cathode or anode.

The titration methods are based on measuring the volume of the solution with the known efficiency of the volumetric solution, which is added in small amounts to the solution of the substance to be determined up to equivalent point, i.e. the condition in which the added volume of the volumetric solution is equal to the theoretical required amount for the reaction with the substance to be determined.

On the basis of the type of reaction:
- acidobasic titration
- reduction–oxidation titration
- titration based on the formation of complexes and compounds with a low degree of dissociation
- precipitation titration

13.3 Polarography

Polarography is an electrochemical method used for the examination of the dependence of electric current on voltage supplied to a couple of electrodes immersed in the electrolysis solution. One of these electrodes is characterised by efficient polarisation, usually it is a mercury droplet electrode, ensuring reproducibility of the measurements. The method is based on the formation of diffusion current whose value is proportional to the content of the substance in the solution to be analysed. The intensity of diffusion current is recorded on a polarogram characterised by the formation of the so-called polarographic waves, whose position characterises the type of substance present in the solution and enables its qualitative determination. Consequently, the height of the wave determines the amount of the substance present in the solution, i.e. quantitative determination. Polarographic analysis has the following advantages:

- it enables the simultaneous determination of several components in the solution, both qualitative and quantitative
- the method is objective, highly accurate, sensitive, was good reproducibility
- a small amount of solution (0.005 ml) is sufficient for analysis
- the consumption of chemical agents is minimal.

13.4. Emission spectral analysis

Emission spectroscopy (ES) includes methods based on the spontaneous emission of particles (atoms) of the analysed substance by the absorption of non-radiation energy (for example, thermal energy) resulting in the transition of these particles to the excited condition. During transition from the excited condition to the condition with a lower energy (for example, the ground condition), the particles emit a characteristic energy quantum which is recorded as an emission line with a specific wavelength. The absolute value of the quantum (the wavelength of the emission line) is given by the qualitative nature of the particles, and the intensity is given by the concentration of the particles. There is atomic emission spectroscopy (AES), based on the emission of radiation of electroneutral atoms or ions, and molecular emission spectrometry (MES), based on the emission of radiation of molecules or particles. AES is divided into spectrometry in the region of optical spectra (vacuum ultraviolet range from 100 to 200 nm, ultraviolet range from 200 to 380 nm, visible range from 380 to 760 nm, and infrared region above 760 nm), and the spectrometry of x-ray radiation, 0.01–10 nm.

Analysis carried out in equipment consisting of an excitation source, a spectrometer for the separation of individual frequencies, a system of lenses or mirrors, providing for the conduction of radiation from the source to the spectral device (detector), used for the measurement of radiation current for the individual frequencies, and finally, if the detector is in the form of a photographic sensitive layer – auxiliary evaluation equipment, projectors, line photometers, microscopes. Depending on the supplied energy, the excitation sources are divided into flame and electrical. In the case of the flame sources, the analysed compound is supplied in the form of aerosol into the flame of gas fuels, combusted in oxygen or in air. The electric source of radiation is the low-temperature plasma of electric discharge excited in arc, spark or high frequency oscillators (direct current arc has a temperature of 5500–6500°C). The spectrometers manufactured commercially by various companies are capable of determining simultaneously up to 20 analysed elements in the matrix, for example, in aluminium, they make it possible to determine the content of the elements Cu, Fe, Si, Mn, Mg, Zn, Ni, Cr, Ti, Pb, Bi, Sn, Be, V, Ca, B, Ga, Zr, Cd.

13.5. Absorption spectrophotometry

Absorption spectrophotometry (AS) is an analytical method based on the absorption of electromagnetic radiation in the visible and ultraviolet ranges of the spectrum. The measured substance absorbs the characteristic quanta of electromagnetic radiation energy, thus increasing the content of the internal energy of the particles of the substance, atoms or molecules. The radiation, not absorbed by the substance, falls on the detector which transforms the radiation energy to electric current. The determined signal must be in a defined relationship to the gradient of current falling on the detector (photoelectric cell) and, also to absorbance A, in order to use the linear relationship of this quantity to concentration C: (Lambert–Beer law of absorption spectroscopy):

$$\log\frac{\Phi^0}{\Phi} = A = \varepsilon C l \tag{92}$$

where Φ^0 is the flux entering into the medium to be measured; Φ is the light flux, leaving the measurement medium (weakened by absorption); A is absorbance; ε is absorption (extinction) factor; C is concentration; l is the thickness of the layer of the measured substance.

Light absorption is measured by photoelectric methods, i.e. the measurement of photoelectric current formed as a result of the incidence of radiation, passed through the absorbing medium, on the detector in the visible or ultraviolet range, or colorimetric methods, i.e. the visual comparison of the colour of the investigated solution with the colour of the solution of the known composition. Equipment for measuring absorption consists of a radiation source usually represented by a deuterium or halogen discharge tube, a monochromator, a filter, a grid and a prism used for the dissociation of polychromatic radiation into a sequence of individual wavelengths, a collimator (forming the beam), measuring cells with the solution to be measured, sensors (photoelectric cells, diodes, multipliers), an amplifier and an evaluation device. Up to 80 elements can be determined by the photoelectric method. The method is fast, sensitive and suitable for the determination of trace concentrations (of the order of parts per million).

13.6. Atomic absorption spectrometry (AAS)

The AAS is based on the examination of the absorption of atomic spectra by free atoms in the gas condition, in most cases in a flame. The Kirchhoff law indicates that the free atoms in the gas condition are capable of absorbing the radiation of the wavelengths identical with those which they themselves emit. In the flame, or using some flame-free technology, for example, an electrically heated measuring cell, the specimen evaporates and is atomised. Subsequently, the gas plasma is subjected to interaction with the emission radiation of the analysed element (the monochromator of the selected emission line of the given specimen) emited by the source, i.e. a hollow cathode. The detector indicates the current of monochromatic radiation weakened by absorption by atoms which is proportional to the concentration of the analysed element in the specimen. Under suitable selected experimental conditions, this method may be used for the quantitative determination of almost all metallic elements of the periodic system of elements. Because of its speed, sensitivity, accuracy and specific features of determination of a large number of elements, atomic absorption spectroscopy is used on an increasing scale in advanced analytical laboratories in many industries.

13.7. Fluorescence spectroscopy

Fluorescence spectroscopy includes the methods based on the measurement of fluorescent radiation emitted by the analysed substance. The common feature of these methods is the primary absorption of energy which increases the total internal energy of the molecules or atoms of the analysed substance and subsequently it is emitted by them as secondary fluorescent radiation. On the basis of the absolute value of the energy quanta of fluorescent radiation, the fluorescence method may be divided into:

- x-ray fluorescence spectrometry. Characteristic radiation in this case is excited by an x-ray lamp or a beam of electrons accelerated by a voltage of approximately 10 kV. Using the method, it is possible to analyse elements with the proton number higher than 11. In the method, measurements are taken of the dependence of the intensity of fluorescent radiation of the substance on the appropriate wavelength, i.e. qualitative analysis, and its concentration, i.e. quantitative analysis. Detectors are represented by scintillation detectors, gas proportional detectors or silicon or germanium semiconductor detectors. Commercially produced analysers contain up to 30 channels, operate in the automatic mode and are very fast. The high output of the x-ray lamps makes it possible to excite the radiation of elements in concentrations of the order of 10^{-3} %. The elements which can be easily analysed include Fe, Mn, Si, Ca, Mg, Al, P, Cr, Ti, Na, K, Ba, Zn, Cu, Ni, V, S, Co, W, Mo, Nb, Ta.

- atomic fluorescence spectrometry (AFS) works in the visible and ultraviolet ranges of the spectrum.
- fluorimetry – the analysed particle is the molecule. The source of ultraviolet radiation is usually a mercury high-pressure discharge lamp.

13.8 Local microanalysis

Local microanalysis is used for examining the chemical composition of microvolumes inside or on the surface. In most cases, it is carried out using spectral analytical methods, such as electron microanalysis, Auger spectrometry, the mass spectrometry of secondary ions, the mass spectrometry of the ion field, electron-induced desorption, etc. The principle of the majority of these methods is that the focused beam of exciting radiation or particles hits the specimen in the analysed area resulting in various physical effects which are subsequently detected and analysed. The exciting beam may be represented by accelerated electrons, ions, protons, and photons of x-ray, visible or ultraviolet radiation. Analysis may be carried out using excited electrons with different energies, characteristic x-radiation, secondary ions, etc.

The electron microanalyser is a vacuum x-ray spectrometer with energy dispersion, supplemented with a scanning electron microscope. The focused electron beam hits the specific area of the specimen and induces characteristic x-radiation there. The method can be used for elements with the proton number of 4 (beryllium) or higher.

Auger electron spectrometry (AES) is the analysis of electrons emitted during the mutual interaction of ionising radiation, most frequently electrons, with the specimen. Auger spectrometry is characterised by the especially high depth resolution (0.5–2.0 nm).

The method is used for examination of microinhomogeneities only in a thin layer on the surface of the specimen.

Mass spectrometry is a method for separating the individual components of the specimen in the form of analysed particles, capable of independent existence in the gas phase. It is the method of two-dimensional separation on the basis of the mass of the components and also their concentration in the specimen. The division process is carried out in electric and magnetic fields of a mass spectrometer. Ionisation of the microvolumes is carried out by, for example, the impact of electrons, high-frequency spark discharge or laser. A photosensitive plate is used for recording the spectral lines, where each line corresponds to a specific isotope of the elements (qualitative analysis). Quantitative analysis is carried out by spectrometry or the method of the last visible line on the photoplate from a series of spectra with gradually decreasing exposure intensity. The mass spectrometry of solids makes it possible to determine up to 75 elements simultaneously.

The mass spectrometry of secondary ions (SIMS method) is characterised by the formation of free ions from solid substances bombarded with primary ions. In the case of accurately determined parameters of the primary ions, secondary particles, characteristic of the overall composition of the surface layers of the investigated specimen, are emitted. This method makes it possible to analyse metallic conducting, semiconductor and non-conducting materials, and can also be used for the determination of the isotope composition. It can be used to determine all elements of the periodic system. The absolute sensitivity of the method is 10^{-16}–10^{-19} g and can be used in the concentration range 1–10%.

13.9. Radiometric methods

The radiometric analytical methods are based on the direct proportionality between the mass of the radioisotope and the intensity of radiation of the isotope. The radioisotope may be present in the analytical substance and, consequently, it is radiometric titration or methods of isotopic dilution. The radioisotope may also be produced artificially in the specimen and this is the method of activation analysis. Another group of methods is based on the measurement of the intensity of external radioactive radiation after its absorption or scattering by the investigated substance.

The radiometric analytical methods are simple and fast, selective, characterised by high sensitivity and are nondestructive. They are suitable for the determination of trace amounts of admixtures in high-purity materials, for example, in semiconductors, in structural materials of nuclear reactors and space materials, and also in biomaterials. In analysis, it is necessary to measure the radiation emitted by the substances or emitted by their radionuclide formed by activation, or measurements are taken of the changes of external radiation (α, β, γ, neutron) after its absorption or backscattering by the analysed substance, or it is necessary to utilise the ionisation caused by radioactive radiation. The detectors of the individual types of radiation include the ionisation chamber, the Geiger–Müller counter, the scintillation probe with a photomultiplier for α particles – ZnS (Ag) scintillator, for β-radiation NaI(Tl), for γ-radiation, the crystal of anthracene and semiconductor detectors Si(Tl), Ge(Tl) with thallium.

In the methods based on the scattering and absorption of radiation, the source of radiation is usually ^{242}Cm (half-life 162 days) or ^{241}Am (458 years); they are both suitable for the determination of lighter elements. γ-emitters are usually represented by radionuclides ^{90}Sr/^{90}Y used for the analysis of binary and multicomponent alloys. Using ^{170}Tm, it is possible to determine the ash content of coal. Neutron analysis is used as a source of fast neutrons ^{241}Am/Be, ^{226}Ra/Be, ^{210}Po/Be.

Radionuclide x-ray fluorescence analysis – this method is based on the excitation and detection of characteristic x-radiation. Suitable emitters are ^{90}Sr/^{90}Y, ^{147}Pm, ^{204}Tl, ^{85}Kr, ^{3}H. The target is usually produced from C, Cu, Ag, Au, Al, Sn. In the determination of lighter elements, good results have been obtained with ^{3}H/Zr and ^{3}H/Ti sources. At present, the x-ray fluorescence analysis systems are produced with two radionuclide sources, for example, with the source ^{239}Pu, which is suitable for the determination of Pb, Cu, Zn, Mn, Cr, Fe, Co, Ni, and with the source ^{109}Cd, for the determination of Mo, Nb, Zr. Radionuclide x-ray fluorescence analysis is used widely in mining, for example, for the inspection of mined ores, in flotation plants, in metallurgy of steel and non-ferrous metals, with a limiting sensitivity of up to 0.001%.

The methods based on the measurement of the natural activity of radioactive isotopes of elements with the long half-time found in nature: K, Rb, In, La, Sm, Re, Ru, Th, U, Rn, Ac, C, Po, Zr, Sn, W, Bi. The measurement of exhalation of radon from bedrock is based on a similar principle.

In radioactivation analysis, the analysed element of the admixture is activated in the same manner as the stable isotopes of the elements present in the specimen for the artificial radioactive isotopes by a nuclear reaction after irradiation with neutrons, protons, deuterons, or high-energy quanta. The resultant radioisotopes differ in the half-life, the type and energy of radiation. This method may be used to not only identify the individual elements but also determine the amount of these elements.

Of greatest importance is neutron activation analysis (NAA). In irradiation with a flux of neutrons, the latter are trapped by atomic nuclei. When trapping a single neutron, the new resultant unstable nucleus has the same sequence number but is heavier by one mass unit. After a specific time of existence (10^{-20}–10^{-12} s), the nucleus is stabilised by the emission of a single photon, proton, alpha particle or two neutrons. Activation is carried out using radioisotope neutron sources, neutron generators, or in most cases a nuclear reactor with a neutron flux of 10^{15}–10^{20} m^{-2} s^{-1} fitted with experimental channels for placing the specimens to be irradiated. If the analysis is carried out without destruction of the specimen, it is instrumental activation analysis (IAA). This method is used widely in analysis of a large number of materials (high-purity metals, rock, biological materials, food, etc), especially in the determination of elements present in trace concentrations in the main compound. By activation in the reactor it is possible to obtain, for radiometric measurements, the radioisotopes of all elements important in metallurgy: Mn (10^{-11}/g), Na, Cu, Al, As, Sb, Ga, V, Ta, W, Au (10^{-10}/g), Cd, Zn, Ba, Ni, Co, P, U (10^{-9}/g), Ag, Hg, Tl, Te, Cr, Mo, Zr (10^{-8}/g). The method can also be used for determination of the lanthanides and elements of the platinum group present in the investigated substance, because they are characterised by large activation efficient cross sections. The radionuclides, formed as a result of radiation, emit in the majority of cases β-radiation with different energies, are characterised by different β and their activity differs by several orders of magnitude.

The concentration gradients (Fig. 32 and 33) and segregation microinhomogeneities, formed during the solidification of metals and indicated by means of radionuclides, are determined by the nondestructive slit method and by the destructive sectioning method [159]. The distribution of the marker atoms is also indicated by autoradiographic examination (Fig. 34b).

13.10. Determination of gases in metals

The content of the gas admixtures or of reaction products with other elements in metals may have a detrimental effect on the properties of materials. In metal melts and also in solid metals, gases as nitrogen, hydrogen, oxygen, and also carbon are present either as interstitial solid solutions or in the atomic or molecular form, as chemical compounds (nonmetallic inclusions), or as absorption or surface compounds. In addition, the metals may contain other gases such as CO, CO_2, CH_4, H_2S, inert gases and water vapours.

The methods of gas analysis may be divided into:

- absorption methods – volumetry, nanometry, electrochemistry, photometry, thermochemistry;
- combustion methods – the determination of the volume changes formed as a result of the combustion of the analysed substance or determination of the calorific effect of the chemical reaction taking place on the solid catalyst (thermometry)
- gas chromatography – the diffusion process taking place between two immiscible phases on the basis of absorption or separation;
- distillation methods – fractional condensation in analytical columns;
- diffusion methods – separation of the isotopes
- measurement of density – the application of gas balance and effusiometers, or the measurement of the velocity of sound and light interference;
- magnetometric methods – determination of the oxygen concentration in gases;
- measurement of the heat conductivity of gases;
- optical methods – absorption methods for the analysis of non-ferrous metals, emission methods for the analysis of inert gases;
- mass spectrometry of gas compounds;
- the activation neutron analysis – the method is used mainly for the determination of oxygen and nitrogen in metals. It is based on the nuclear reaction ^{16}O (n,p) ^{16}N. In most cases, the detector is in the form of a scintillation crystal NaI(Tl). The measurement time is 20–30 sec, the limiting sensitivity is up to 10^{-5}% of oxygen. In the case of nitrogen, the irradiation and measurement time is approximately 10 min, sensitivity is 10 ppm of nitrogen.

13.11. Evaluation of analytical methods of determination of pure metals

The most universal method of analysis of metals is spark mass spectrometry. In analysis, it is possible to determine 70–75 elements (admixtures), with the limiting sensitivity of 10^{-6}–10^{-8} wt.%, with the consumption of several milligrams of the analysed substance. However, quantitative determination requires standards which in many cases are not available.

In most cases, the analytical control of the quality of pure metals is carried out using spectral and chemical methods. The advantage of these methods is the possibility of the joint determination of a relatively large number of 15–20 elements with high sensitivity and a simple analysis procedure. However, the spectral analysis of refractory metals Ta, Nb, Zr, Hf is far more difficult than analysis of In, Ga, Sb. The problem is that the optical spectra of the refractory metals are very complicated. The sensitivity limit for the admixtures Fe, Ni, Si in Sb is on the level of 10^{-5} wt.%, whereas in niobium for the same admixtures it is only 10^{-3}–10^{-4}%.

Activation analysis is one of the methods with high sensitivity. Thermal neutrons are utilised for the determination of elements in the middle part of the periodic system of elements. A suitable example of the efficient application of the possibilities of this method is the analysis of niobium and titanium characterised by isotopes with a short half-life and it is therefore possible to analyse specimens weighing 1–2 g immediately after irradiation. Achieving the limiting sensitivity of analysis is disrupted by the presence of admixtures Ta and W characterised by the formation of isotopes with a long half-life.

The sensitivity of analytical methods may be efficiently increased by several orders of magnitude by the method of concentration of admixtures. In this method, a physical–chemical procedure is used to separate the main metal – matrix. Therefore, the main direction of the development of analytical methods for pure metals is the development of an efficient method of concentration of the admixtures together with the optimum analysis method. The main metal is separated out using a method similar to those in the refining of metals, i.e. extraction, distillation, sublimation, selective precipitation, reduction. From the analytical methods, it is recommended to use emission spectrometry, polarography or atomic absorption photometry.

An important direction in the development of analytical procedures is the construction of measuring systems and new highly sensitive methods for the determination of the concentration of the admixtures oxygen, nitrogen, hydrogen and carbon in metals. Equipment with a mass spectrum analyser, based on vacuum extraction, makes it possible to determine oxygen content of up to 2 ppm, and in the case of hydrogen and nitrogen up to 0.5 ppm, for carbon up to 0.20 ppm, and the duration of single analysis is approximately 5 min. High sensitivity (10^{-6}%) may be achieved using gamma activation analysis combined with vacuum extraction. A linear accelerator is used as a radiation source. After irradiating the specimen, removing the surface layer, and separating the admixtures to be determined, it is necessary to measure the activity of elements directly in the gas phase.

The main assumption for the applicability of a specific analytical method is, in particular, good reproducibility of the results. It is always important to consider the aim for which the given method should be used. An important question is that of standards, calibration, equipment facilities and the possibility of occurrence of errors in the preparation, analysis or evaluation of samples.

13.12. Resistance of metals and semiconductors

Electrical resistance and resistivity are important characteristics of the purity of materials and they are determined by the composition, physical properties and structural inhomogeneity of metals, alloys, elementary semiconductors or their compounds. The determination of resistivity provides important information on the given substance, its electronic structure, phase changes, or the type of magnetic ordering. An advantage of the method is the relatively easy measurement procedure of resistivity and a high relative accuracy of measurements which reaches 10^{-6}. However, only the total effect of the admixtures on the properties of the material is determined.

In semiconductor materials, where the requirements on purity are extremely high and the total content of admixtures is usually lower than 10 ppm, it is no longer possible to use direct analytical methods for the determination of the presence of admixtures. In addition to this, intentional microalloying of semiconductors results in the electronic or hole conductivity, and it is therefore necessary to consider primarily the concentration and distribution of charge carriers (homogeneity of the crystal), their mobility and life-time of minor carriers. The best results in these applications have been obtained using the methods of measuring resistivity ρ_p (or conductivity σ_p) of semiconductors, measurement of Hall voltage U_H or Hall constant R_H for the determination of the type and concentration of current carriers in semiconductors.

In most cases, the resistivity of semiconductors is measured using the compensation four- or two-point method. Resistivity is determined from the magnitude of the current passing through the semiconductor, the geometrical dimensions of the specimen and from the voltage drop between the measuring electrodes spaced at a specific distance. The resultant measured value is used for the direct calculation of the concentration of major current carriers using the equation ($C^*_{maj} \gg C^*_{min}$):

$$\rho_p = \frac{1}{C^*_{maj} e \mu_{maj}} \tag{93}$$

where $C^*_{maj}(C^*_{min})$ is the concentration of the major (minor) carriers of the current of electrons in semiconductors of type n, holes of type p [cm^{-3}]; μ_{maj} is the mobility of the major carriers [$cm^2\ V^{-1}\ s^{-1}$]; e is the electron charge ($1.602 \cdot 10^{-19}$ C).

In the evaluation of the purity and perfection of the structure of metallic materials, it is recommended to use the method of measuring residual resistivity ρ_0 at a temperature of 0 K, or the ratio of the resistivity at room temperature $\rho_{295\ K}$ and the liquid helium temperature $\rho_{4.2\ K}$. Residual resistivity ρ_0, which appears in all compounds, is caused by crystal lattice defects (vacancies, dislocations...) and by the presence of admixtures and undesirable elements. The value of ρ_0 is also affected by the structural inhomogeneity of the sample, internal stresses, etc.

The method is based on the Matthiessen law according to which electrical resistance consists of two components: the component dependent on the scattering of electrons by thermal oscillations of the atoms, and the component determined by the scattering of electrons on admixtures and lattice defects. Because the thermal scattering of electrons at the liquid helium temperature is small, the value of the electrical resistance of the crystal is in fact determined only by the second component. Therefore, the ratio of the resistivity $\Gamma = \rho_{295\ K}/\rho_{4.2\ K}$ may be used approximately as the measure of the total purity and structural perfection of metals. The value of Γ is usually expressed in terms of commercial purity metals, in hundreds or thousands for pure metals and several tens of thousands for superpure metals, Table 16 [48,105].

At present, the measurements of the resistivity of metals are taken in most cases by means of the contact potentiometric or contactless method. Disadvantages of the contact four-point method include high sensitivity to the experiment conditions, the quality of brazed contacts, the shape and external defects in the sample. The effect of superconducting admixtures and compounds in the main metal is also negative. Nevertheless, the method is used widely in the complex and accurate rapid determination of the quality of crystals and single crystals and of their relative purity in relation to standard specimens. In the contactless method, stringent requirements are imposed on the equipment and devices. The method of alternating magnetisation is based on the change of the impedance of the coil by inserting a metallic core. The specimen to be measured is inserted into the magnetic field, generated by a coil powered with a current of specific frequency.

Table 16. Values of $\Gamma = \rho_{295\,K}/\rho_{4.2\,K}$ for several refractory metals after electron beam zone refining in a vacuum of 10^{-3} Pa, $v = 5 \cdot 10^{-3}$ cm s^{-1}

Element	Origin of specimens	Specimen diameter [mm]	Values of Γ for the number of passes of the zone *n*				
			0	1	2	3	5
Tungsten	TR	3.5	2200	10600	26000		
	SV	4	800	13750	21000	39000	
	Literature	6	40	8900		18300	20000
Molybdenum	TR	6	31	123	150	230	2780
	Literature	8	50	60		740	3000
Tantalum	SV	4	9	16		55	181
	Literature			27		39	78
	Literature	3	104			340	552
Niobium	PL	5.5	350		2650		

TR – Tesla Roznov, SV – Safina Vestec, Czech Republic, PL – Plansee, Austria

Currents are induced in the specimen and influence the main excitation current of the coil. This is reflected in a change in the impedance of the coil in relation to the values measured without the specimen. The magnitude of this change depends on the resistivity and geometrical dimensions of the specimen.

These methods are often used for the determination of the total purity, especially in refractory metals, for the examination of the refining effect of solidification processes, and for the examination of the axial and radial homogeneity and structural perfection of crystals and single crystals.

13.13. Methods of examination of the structure and perfection of crystals

A number of methods are used for the examination of the structure, crystallographic orientation and perfection of metallic crystals:

- metallographic examination of the macro- and microstructure of crystals by optical microscopy;
- electron microscopy using emission or scanning electron microscopes;
- x-ray diffraction for the determination of crystal orientation;
- x-ray topography for the examination of the structure and perfection of crystals.

Optical metallographic microscopes (resolution up to 0.2 μm) may be used for the examination of the following main details of the crystal structure: the shape and dimensions of grains, the dendritic or other inhomogeneity, the shape and nature of distribution of etched images, inclusions, twins; microhardness tests can be carried out. Using special thermal chambers, it is possible to examine directly the processes of recrystallisation and phase transformations, crystallisation processes, grain growth, etc. There are a large number of etching procedures which can be used to develop the grain and subgrain boundaries, dendritic, cellular or other microstructures or substructures, growth of dislocations, and the dislocations formed during plastic deformation. The colour etching method, based on the electrolytic oxidation of the surface of the section in special solutions or by of ion etching, is used for the identification of the orientation of grains and textures or nonmetallic inclusions in crystals.

In electron microscopes, light radiation is replaced by an electron beam which falls on and partially penetrates into the specimen. This results in the scatter and refraction of electrons or electrons may be reflected. The resolution power of transmission electron microscopes at an accelerating voltage from 100 kV to 1 MV is up to 0.35 nm. Replicas (imprints or thin foils) are used as preparations. These systems are used in the examination of various defects, dislocations, inclusions or precipitates. In scanning electron microscopes, a narrow electron beam travels in rows on the specimen. After the electrons impacted the specimen, they are reflected or secondary electron or x-radiation forms. The main advantage of scanning electron microscopes is that they require almost no preparation of specimens and enable sharp images of the surface of examined object to be obtained at a large depth. They may be supplemented by an x-ray spectrograph for examination of the distribution of the individual elements in the specimen or an electron microanalyser.

The optical method is used for the approximate determination of the orientation of single crystals and bicrystals. This method is based on the reflection of the light beam from etches planes on the surface of the crystal. Selective etching procedures may be used to produce suitable etch pits which are reflected on the screen in the form of typical reflection images. From the shape and sequence of the resultant inflection images during rotation of the specimen around its axis and from the angles of two crystal planes it is possible to determine the orientation of the crystal and the crystallographic deviation of orientation from the crystal axis quite rapidly, without destruction of the specimen and with a reproducible accuracy of $<2°$.

X-ray diffraction is based on the bending or diffraction of x-radiation during its impact on the examined metal and makes it possible to determine the arrangement of the atoms in the metal lattice and determine the lattice parameters with a relatively high accuracy. According to the method used, monochromatic or polychromatic x-radiation is used. The main methods of electron diffraction are:

- Laue method – determination of the crystal axes in metallic crystals
- back reflection method – determination of the lattice parameters from the Bragg equation, internal stress, deformation of the metal lattice, crystal orientation.
- Debye–Scherrer method – determination of the type of lattice in powder materials and polycrystalline metals, differentiation of the crystal and amorphous structure, identification of different types of crystals, qualitative and quantitative analysis.

The direct methods of examination of crystal defects include x-ray diffraction topography used for the examination of the dislocation structure of semiconductor crystals with a low dislocation density. The Berg–Barrett method makes it possible to determine the size of the subgrains of single crystals and their angular disorientation, examine plastic deformation and stability of the crystal structure. In this method, it is possible to obtain a very accurate image from the viewpoint of perfection of the structure as a result of the reflection of x-radiation from the polished surface of the single crystal. At characteristic irradiation, a topogram is produced from the entire surface by discrete rotation of the specimen along the axis situated on the surface of the section. On the basis of the width dark and light areas on the topogram and the magnitude of their shift, it is possible to determine the angles of disorientation of two adjacent grains. The method also makes it possible to detect different inclusions (for example, carbides) which form as bright spot on the topogram. The Lang method is used for determination of mean dislocation density (for example, in silicon). This method determines imperfections in the crystal structure as a result of the scatter of x-radiation on lattice defects. Point defects – vacancies and interstitial atoms are studied using an ion projector, magnification 10^6 to 10^7, resolution 0.3 nm.

This list does not cover all the possible methods of the determination of the physical and physical–chemical properties of substances. The methods of determination of the properties are efficiently combined in order to accelerate and improve the accuracy of determination by computing which is also capable of carrying out a number of other statistical or correlation operations, especially in connection with comprehensive data banks of comparison data.

References

[1] Chaudron G.: *Monographies sur les métaux de haute pureté*. Masson, Paris, T.1 – 1972, 540 p., T.2 – 1977, 644 p. T.3 – 1977, 714 p., in French.

[2] Chalmers, B.: *Physical metallurgy*. New York, London. 1962, 455 p.

[3] Belyaev A.I.: *Physico-chemical principles of refining of metals and semiconductors*. Ed. Metallurgiya, Moscow, 1973, 222 p., in Russian.

[4] Dimitrov O.: *Les Métaux de haute pureté*. Presses Universitaires de France, Paris, 1978, 178 p., in French.

[5] Pfann, W.G.: *Zone melting*. New York, London, 1958, 366 p.

[6] Barthel, J., Buhrig, E., Hein, K. and Kuchař, L.: *Kristallisation aus Schmelzen*. Leipzig, VEB Deutscher Verlag für Grundstoffindustrie, 1983, 356 p., in German. Spravochnik "*Kristallizacija iz rasplavov*". Moscow, Ed. Metallurgiya, 1987, 320 p., in Russian.

[7] Kuchař, L.: *Metallurgy of pure metals. Part 1. Crystallization processes*. Textbook of theVŠB-TU of Ostrava, 1. Ed., 1988, 2. Ed. 1992, 338 p., in Czech.

[8] Wilke, K.-Th.: *Kristallzüchtung*. VEB Deutscher Verlag der Wissenschaften, Berlin, 1973, 924 p., in German.

[9] Hurle, D.T.J.: *Handbook of crystal growth*. vol. 1, 2, 3. North-Holland, 1994.

[10] Massalski T.B.: *Binary Alloy Phase Diagrams*. American Soc. for Metals, Metals Park, Ohio, 1986, 2224 p.

[11] Dean J. A.: *Chemical separation methods*. SNTL, Prague, 1974, 401 p., in Czech.

[12] Juračka F.: *Ion enchangers*. SNTL, Prague, 1965, 150 p., in Czech.

[13] Grígerová, T., Lukáč, I. and Kořený, R.: *Casting of non-ferrous metals*. Alfa Bratislava, 1998, 421 p., in Slovak.

[14] Feuerbacher, B., Hamacher, H. and Naumann, R.J.: *Materials sciences in space*. Springer Verlag, Berlin, 1986, p. 164.

[15] Kaufman. L. and Bernstein, H.: *Computer calculation of phase diagrams*. Academic Press, New York - London, 1970, 326 p.

[16] Hultgren, R. et al.: *Selected values of the thermodynamic properties of binary alloys*. American Society for Metals, Ohio, 1973, 1436 p.

[17] Alper, A.A.: *Phase diagrams, materials science and technology*. Academic Press, London, 1970, 358 p.

[18] Drápala, J., Byczanski, P. and Kuchař, L.: Computer calculation of equilibrium solidus and liquidus curves, *Transactions of the VŠB-Technical University of Ostrava, Metallurgical Series,* 30, 1984, No. 1, p. 73–79, in Czech.

[19] Vřešťál, J. and Kuchař, L.: Addition distribution thermodynamics in liquid and solid phases of alloys. *Hutnické listy,* vol. 45, 1990, No. 4, p. 286–292, in Czech.

[20] Hayes, A. and Chipman, J.: *Trans. AIME,* 135, 1939, p. 85.

[21] Hansen M.: *Constitution of Binary Alloys*. New York, 1958.

[22] Elliott R.P.: *Constitution of Binary Alloys*. New York, 1965.

[23] Shunk F.A.: *Constitution of Binary Alloys*. New York, 1969.

[24] Drápala, J. and Kuchař, L.: *Metallurgy of pure metals. Exercise instructions.* Textbook of the VŠB-TU of Ostrava, 1990, 166 p., in Czech.

[25] Kuchař, L., Kuchař, L. jr. and Drápala, J.: Contribution of copper refining by crystallization methods. *Acta Metallurgica Slovaca*, vol. 4, Special Issue, 1998, No, 3, p. 101–109, in Czech.

[26] Alexandrov B.N., Rybalchenko, L.F. and Dukin, V.V.: *IAN SSSR, Metally*, 1970, No. 4, p. 70–75, in Russian.

[27] Vigdorovich, V.N., Marychev, V.V. and Shirobokova, T.G.: *IAN SSSR, Metally*, 1970, No. 2, p. 129–134, in Russian.

[28] Tolmie, E.D. and Robins, D.A.: *J. Inst. Metals*, 85, 1957, No. 5, p. 171.

[29] Gerlach J., Powlek F. und Rogala, D.: Zonengeschmolzenen Kupfer. *Metall*, 1964, No. 11, p. 1158, in German.

[30] Kuchař, L.: Importance of distribution coefficients of admixtures in metals for segregations and refining processes. In: *Metal 94*, Ostrava, II, p. 1–19, in Czech.

[31] Drápala, J., Kuchař, L. and Burchanov, G.S.: Distribution coefficients of impurities in metals: Periodic dependence on the atomic number of impurity. *Inorganic materials*, vol. 34, 1998, No. 2, p. 114–127.

[32] Kuchař, L.: Verteilungskoeffizienten beim Zonenschmelzen von Aluminium. *Bergakademie*, 14, 1962, 9, p. 622–631, in German.

[33] Kuchař, L.: Segregation coefficients in aluminium. *Hutnické listy*, XVII, 1962, 9, p. 639–648, in Czech.

[34] Vigdorovich, V.N.: *Refining of metals and semiconductors by crystallization.* Metallurgiya, Moscow, 1969, p. 39, in Russian.

[35] Burton J.A., Prim R.C. and Slichter W.R.: *Journ. Chem. Phys.*, 21, 1953, p. 1987.

[36] Tiller, W.A. and Rutter, J.W.: *Canad. Journ. Phys.*, 34, 1956, p. 96.

[37] Kuchař, L., Drápala, J., Duží, P., Ivanov, J.M., Kostava, N.I. and Dvinin, S.J.: *Equipment for refining of volatile substances.* Patent No. 276366, 20.3.1992, Prague, in Czech.

[38] Burris, L., Stockman, C.H. and Dillon, J.G.: *Trans. AIME*, 1955, p. 1017–1023.

[39] Kuchař, L.: Equilibrium and effective distribution coefficients of the ^{95}Zr radionuclide in zone melted aluminium. *Metallic materials*, vol. VI, 1968, No. 6, p. 571–578, in Czech.

[40] Gorelik, S.S. and Dashevski, M.J.: *Material science of semiconductors and dielectrics*. Ed. Metallurgiya, Moscow, 1988, p. 208, in Russian.

[41] *Smithells Metals Reference Book.* Six edition, Butterworths, London, 1983.

[42] Dembovský, V.: *Plasma metallurgy*, SNTL, Prague, 1978, 260 p., in Czech.

[43] Kuchař, L.: *Metallurgy of non-ferrous metals. Part I. Pyrometallurgy*. ES VŠB Ostrava, 1987, 335 p., in Czech.

[44] Kuchař, L. and Drápala, J.: Manufacture of high-purity metals by crystallization methods. *Hutnické listy*, vol. 40, 1985, No. 7, p. 498–507, in Czech.

[45] Ackermann, G. et al.: *Elektrolytgleichgewichte und Elektrochemie*. Lehrbuch 5, Lehrwerk Chemie, Leipzig, 1988, 292 p.

[46] Belashczenko D. K. and Grigoryev G. A.: *Izv. VUZ, Chernaya Metallurgiya*, 1962, No. 1, p. 124.

[47] Tomík, B., Poljak B. and Jirkovský R.: *Analytical chemistry for metallurgists*. SNTL/Alfa, Prague, 1988, 261 p., in Czech.

[48] Drápala, J.: *Preparation of single crystals of refractory metals with defined purity and properties*. Thesis, VŠB–TU of Ostrava. 1985, 217 p., in Czech.

[49] Kuchař, L., Wozniaková, B. and Drápala, J.: Segregation behaviour during the zone melting of refractory metals. *Journal of Crystal Growth*, vol. 52, 1981, p. 359–366.

[50] Drápala, J., Kuchař, L. and Kursa, M.: Preparation of high purity metals by crystallization methods. *Journal de physique IV*, Colloque C7, vol. 5, 1995, p. C7–143 – C7–148.

[51] Kuchař, L. and Wozniaková, B.: Distribution coefficients of admixtures in silber. *Transactions of the VŠB-Technical University of Ostrava, Metallurgical Series*, XXVIII, 1982, No. 1, p. 1–9, in Czech.

[52] Kuchař, L., Drápala, J. and Wozniaková, B.: Distribution coefficients of admixtures in copper, silber and gold. *Metallic materials*, 21, 1983, No.1, p. 80–89, in Czech.

[53] Kuchař, L. and Revel, G: Coefficient de partage des impurités présentes dans l'aluminium, *Comptes Rendus Acad. Sc. Paris*, t.262, 1966, serie C, p. 1365–1367, in French.

[54] Kuchař, L.: Deplacement du zirconium ^{95}Zr au cours de la zone fondue de l'aluminium. *Comptes Rendus Acad. Sc. Paris*, t.262, 1966, serie C, p. 1466–1469, in French.

[55] Kuchař, L.: Distribution coefficients of admixtures by zone melting of aluminium. *Zone melting.* In: Proceedings, Ed. V.N.Vigdorovich, Metallurgiya, Moscow, 1966, p. 145–158, In Russian.

[56] Kuchař, L.: A zónásau olvasztott alumínium megoszlási együtthatói. *Kohászati Lapok*, 1964, No. 2, p. 65–69, in Hungarian.

[57] Kuchař, L.: Une méthode de détermination des coefficients de partage théoriques. Aplication au systéme aluminium–yttrium. *Comptes Rendus Acad. Sc. Paris*, t.262, 1966, serie C, p. 393–396, in French.

[58] Kuchař, L. and Czoboly, E.: Computational and empirical determination of distribution coefficient of zinc in aluminium. *Transactions of the VŠB–Technical University of Ostrava, Metallurgical Series*, XI, 1965, No. 5, p. 771–784, in Czech.

[59] Kuchař, L.: New method of the visualization of binary diagrams of alloys and determination of distributing coefficients applied in aluminium–yttrium system. *Transactions of the VŠB-Technical University of Ostrava, Metallurgical Series*, XIII, 1967, No. 1, p. 31–40, in Czech.

[60] Kuchař, L. and Revel, G.: Contribution to the theory of the preparation of high-purity aluminium by zone melting. *Transactions of the VŠB-Technical University of Ostrava, Metallurgical Series*, XIII, 1967, special issue, p. 3–18, in Czech.

[61] Kuchař, L. and Habrman, P.: Verteilung von ^{55}Fe und ^{110m}Ag beim horizontalen Breitbandgiessen von Aluminium. *Transactions of the VŠB-Technical University of Ostrava, Metallurgical Series*, XXXIX, 1993, No. 1, p. 27–34, in German.

[62] Kuchař, L., Drápala, J. and Kuchař jr., L. Distribution of admixtures in aluminium alloys and calculation of solidification interval. *Transactions of the Technical University of Košice*, 1999, no. 3, p. 38–45, in Czech.

[63] Drápala, J., Kuchař, L. and Wozniaková, B.: Distribution coefficients of admixtures in gold. *Transactions of the VŠB-Technical University of Ostrava, Metallurgical Series*, XXX, 1984, No. 1, p. 53–60, in Czech.

[64] Kuchař, L., Drápala, J. and Luňáček, J.: Purification methods of Cd, Te and CdTe and periodicity of distribution coefficients of admixtures. *Journal of Crystal Growth*, 161, 1996, p. 94–103.

[65] Kuchař, L., Wozniaková, B., Duží, P. and Drápala, J.: Distributing of ^{124}Sb and ^{110}Ag in iron, cobalt and copper during the arc zone melting. *Transactions of the VŠB-Technical University of Ostrava, Metallurgical Series*, XXIV, 1978, No. 2, p. 81–86, in Czech.

[66] Kuchař, L., Drápala, J. and Kuchař, L. jr.: Behaviour of admixtures by crystallization of chromium, vanadium and rhenium. In: *High-Purity Materials in Science and Technology*, Dresden, 1985, p. 49–50.

[67] Kuchař, L. and Buhrig, E.: Distributing of antimony ^{124}Sb in the zone melting of copper. *Transactions of the VŠB-Technical University of Ostrava, Metallurgical Series*, XVII, 1971, No. 2, p. 1–13, in Czech.

[68] Kuchař, L. and Tac Le Cao: Equilibrium distribution coefficients of copper–admixture binary systems. *Transactions of the VŠB-Technical University of Ostrava, Metallurgical Series*, XVIII, 1972, No. 3, p. 65–76, in Czech.

[69] Kuchař, L. and Repiská, L.: Method of the calculation of the distribution coefficient of phosphorus in iron for purposes of zone melting and refining. *Transactions of the VŠB-Technical University of Ostrava, Metallurgical Series*, XI, 1965, No. 5, p. 785–794, in Czech.

[70] Repiská, L. and Kuchař, L.: Distribution coefficients in zone refining of iron. The application to Fe–C, Fe–Be systems. *Transactions of the VŠB-Technical University of Ostrava, Metallurgical Series*, XVI, 1970, No. 1, p. 25–44, in Slovak.

[71] Kuchař, L. and Balušek, M.: Preparation of pure iron by arc zone melting. *Transactions of the VŠB-Technical University of Ostrava, Metallurgical Series*, XVI, 1970, No. 1, p. 45–59, in Czech.

[72] Kuchař, L., Repiská, L., Komorová, L. and Baffyová, Z.: Computation of the solidus and liquidus curves in the iron–admixture binary diagrams in the range of low concentrations of admixtures. *Transactions of the VŠB-Technical University of Ostrava, Metallurgical Series*, XXVI, 1980, No. 1, p. 119–136, in Czech.

[73] Repiská, L., Kuchař, L. and Baffyová, Z.: Distribution of admixtures in the region of iron phase tranformations. *Metallic materials*, vol. 19, 1981, No. 5, p. 552–560 in Slovak.

[74] Baffyová, Z., Repiská, L. and Kuchař, L.: Studying the Fe–Sb binary diagram and distribution of ^{124}Sb isotope in iron. *Transactions of the VŠB-Technical University of Ostrava, Metallurgical Series*, XXX, 1984, No. 2, p. 1–12, in Czech.

[75] Kuchař, L. and Repiská, L.: Studying the iron – admixture system in the region of transformations $Fe_{\delta-\gamma}$ and $Fe_{\gamma-\alpha}$. *Transactions of the VŠB-Technical*

University of Ostrava, Metallurgical Series, XXX, 1984, No. 2, p. 35–53, in Czech.

[76] Drápala, J.: Zachowanie sie domieszek przy krystalizacji żelaza. *Inzynieria materialowa*, XV, 1994, No. 2, p. 29–33, in Polish.

[77] Kuchař, L., Buhrig, E. and Duží, P.: Modelling of the solidus and liquidus curves in the GaAs–Ga system. *Transactions of the VŠB-Technical University of Ostrava, Metallurgical Series*, XXXI, 1985, No. 1, p. 15–27, in Czech.

[78] Drápala, J., Buhrig, E., Kuchař, L. and John, M.: On concentration undercooling in the GaAs–Ga system. *Transactions of the VŠB-Technical University of Ostrava, Metallurgical Series*, XXXII, 1986, No. 1, p. 191–202, in Czech.

[79] Chmelařová, M., Drápala, J. and Kuchař, L.: Study of distribution coefficients of admixtures in gallium arsenide. *Transactions of the VŠB-Technical University of Ostrava, Metallurgical Series*, XXXVI, 1990, No. 1, p. 107–127, in Czech.

[80] Buhrig, E., Kuchař, L., Duží, P. und Hein, K.: Beitrag zür Systematisierung der Solidus- und Liquiduskurven und Verteilungskoeffizienten der Systeme Germanium – Verunreinigung. *Transactions of the VŠB-Technical University of Ostrava, Metallurgical Series*, XXVIII, 1982, No. 1, p. 59–82, in German.

[81] Kuchař, L.: Das Verhalten von Verunreinigungen bei der Kristallisation von Germanium. In: 100 Jahre Germanium, Freiberger Forschungshefte, A767, DVG Leipzig, 1987, p. 53–72, in German.

[82] Luňáček, J. and Kuchař, L.: Distribution coefficients of admixtures in silicon and germanium and their relations to generalized moment of admixtures. *Transactions of the VŠB-Technical University of Ostrava, Mining and geological series*, XXXVII, 1991, No. 2, p. 223–238.

[83] Luňáček, J. and Kuchař, L.: Correlation of distribution coefficients of admixtures in germanium and silicon on the generalized moment. *Metallic materials*, vol. 30, 1992, No. 5, p. 393–403, in Czech.

[84] Drápala, J. and Kuchař, L.: Distribution coefficients in semiconductors. *Journal of Electrical Engineering*, vol. 50, 1999, No. 2/s, p. 11–15.

[85] Kuchař, L. and Wozniaková, B.: Periodical dependence of distribution coefficients of admixtures in hafnium. *Transactions of the VŠB-Technical University of Ostrava, Metallurgical Series*, XXIV, 1978, No. 2, p. 19–24, in Czech.

[86] Ojha, V. and Kuchař, L.: Calculation of binary diagrams from the thermodynamic data for which the component behaviour is ideal. *Transactions of the VŠB-Technical University of Ostrava, Metallurgical Series*, XXXII, 1986, No. 1, p. 83–96, in Czech.

[87] Branžovský, J., Drápala, J. and Kuchař, L.: Equilibrium distribution coefficients of admixtures in indium. *Transactions of the VŠB-Technical University of Ostrava, Metallurgical Series*, XXX, 1984, No. 2, p. 19–26, in Czech.

[88] Branžovský, J. and Kuchař, L.: Determination of distribution coefficient of silver in indium. *Transactions of the VŠB-Technical University of Ostrava, Metallurgical Series*, XXX, 1984, No. 2, p. 27–34, in Czech.

[89] Kuchař, L., Drápala, J. and Branžovský, J.: Bestimmung von Verteilungskoeffizienten in Indium mittels radioaktiver Nuklide. *Isotopenpraxis*, vol. 20, 1984, No. 6, p. 207–210, in German.

[90] Drápala, J. and Kuchař, L.: Refining of precious metals by crystallization methods. *Metallic materials*, 35, 1997, No. 3, p. 162–173, in Czech.

[91] Kuchař, L. jr., Kuchař, L. and Drápala, J.: Distribution coefficients of admixtures in gadolinium. In: Proceedings of the 4th Intern. Conference of the VŠB Ostrava, section: Metallurgy of non-ferrous metals and nuclear metallurgy, Ostrava, 1990, p. 22, in Czech.

[92] Kuchař, L. and Kuchař, L. jr.: Distribution of admixtures by zone refining of lanthanides *Transactions of the Technical University of Košice*, vol. 2, 1992, No. 1, p. 89–99 in Czech.

[93] Kuchař, L.jr., Kuchař, L., Drápala, J., Burkhanov, G.S. and Chistyakov, O.D.: The distribution coefficients of admixtures and dysprosium in gadolinium. *Transactions of the VŠB-Technical University of Ostrava, Metallurgical Series*, XXXVIII, 1992, No. 1, p.19–27, in Czech.

[94] Kuchař, L. and Kuchař, L.jr.: Distribution of admixtures by zone refining of lanthanum. *Hutnické listy*, 47, 1992, No. 11, s 24–30, in Czech.

[95] Kuchař, L. and Kuchař, L.jr.: Periodical dependence of distribution coefficients of admixtures in rare-earth metals. *Metallurgy and Foundry Engineering*, AGH Cracow, 19, 1993, No. 1, p. 81–98.

[96] Kuchař, L. and Drápala, J.: Systematic study of distribution coefficients of admixtures in rare-earth metals during crystallization and refining processes. *International Journal of Materials & Product Technology*. vol. 8, 1993, No. 1, p. 58–70.

[97] Kuchař, L.jr., Kuchař, L., Drápala, J., Burkhanov, G.S. and Chistyakov, O.D.: The distribution coefficients of admixtures in gadolinium. *International Journal of Materials & Product Technology*, vol. 9, 1994, No. 4/5/6, p. 323–330.

[98] Kuchař, L., Drápala, J., and Kuchař, L.jr.: Distribution coefficients of admixtures in rare-earth metals. *Vysokochistye veshchestva*, No. 2, 1996, p. 50–58, in Russian.

[99] Kuchař, L., Duží, P. and Wozniaková, B.: Gleichgewichtsverteilungskoeffizienten in Molybdän und Niob, *Neue Hütte*, vol. 21, 1976, No. 5, p. 297–300, in German.

[100] Drápala, J. and Kuchař, L.: Study of refining grade of the electron beam zone melted refractory metals W, Mo and Nb by the help of mass spectral analysis. *Transactions of the VŠB-Technical University of Ostrava, Metallurgical Series*, XXIV, 1978, No. 1, p. 1–13, in Czech.

[101] Kuchař, L., Duží, P., Drápala, J., Barthel, J., Eichler, K., Jurisch, M., Löser, W., Ovsijenko, D.E. and Sosnina, E.J.: Makrosegregation, Mikrosegregation und Substruktur in zonengeschmolzenem Molybdän. *Transactions of the VŠB-Technical University of Ostrava, Metallurgical Series*, Teil I, II, III, XXX, 1984, No. 1, p. 17–52, in German.

[102] Wozniaková, B. and Kuchař, L.: Distribution coefficients of admixtures in niobium. *Transactions of the VŠB-Technical University of Ostrava, Metallurgical Series*, XXIII, 1977, No. 1, p. 1–9, in Czech.

[103] Kuchař, L. and Wozniaková, B.: Verteilung von Iridium im zonengeschmolzenen Niob. *Neue Hütte*, vol. 23, 1978, No. 6, p. 212–214, in German.

[104] Kuchař, L., Wozniaková, B. and Drápala, J.: Determination of distribution coefficient of Ir, Ta and W in niobium and Ta in tungsten by the help of activation analysis. *Transactions of the VŠB-Technical University of Ostrava, Metallurgical Series*, XXIV, 1978, No. 2, p. 59–65, in Czech.

[105] Drápala, J., Cigáň, A. and Kuchař, L.: High purity niobium for superconductor electronics. *Transactions of the VŠB-Technical University of Ostrava, Mining and geological series*, XXXVII, 1991, No. 2, p. 239–256, in Czech.

[106] Kuchař, L. and Dimitrov, O.: Ségragation du cuivre et de l or au cours de la zone fondue du nickel en nacelle de cuivre refroidie. *Comptes Rendus Acad. Sc. Paris*, t. 262, 1966, serie C, p. 718–721, in French.

[107] Kuchař, L., Deschamps, N., Wache, A., Dubois, B. and Dimitrov, O.: Purification du nickel par fusion de zone sur nacelle métallique refroidie. *Les Mémoires Scientifiques de la Revue de Métallurgie*, vol. 64, 1967, No. 9, p. 785–794, in French.

[108] Dimitrov, O. and Kuchař, L.: On the zone melting of nickel. *Transactions of the VŠB-Technical University of Ostrava, Metallurgical Series*, XVI, 1970, No. 1, p. 1–24, in Czech.

[109] Wozniaková, B. and Kuchař, L.: Distribution coefficients of admixtures in nickel. *Annales de chimie – France, science des matériaux*, 1981, No. 6, p. 301–307.

[110] Drápala, J. and Kuchař, L.: Distributing of admixtures in nickel during crystallization. *Transactions of the VŠB-Technical University of Ostrava, Metallurgical Series*, XXX, 1984, No. 2, p. 55–66, in Czech.

[111] Krištofová, D., Kuchař, L. and Kárník, T.: Distribution coefficients of admixtures in lead and their significance for zone refining. *Transactions of the VŠB-Technical University of Ostrava, Metallurgical Series*, XXXVI, 1990, No. 1, p. 167–189, in Czech.

[112] Kuchař, L. and Drápala, J.: To the theory of palladium refining by means of crystallisation procedures. *Hutnické listy*, 51, 1996, No. 12, p. 21–25, in Czech.

[113] Drápala, J. and Kuchař, L.: Distribution coefficients of admixtures in palladium. *Transactions of the VŠB-Technical University of Ostrava, Metallurgical Series*, 42, 1996, No. 1, p.27–35, in Czech.

[114] Drápala, J. and Kuchař, L.: Distribution coefficients of admixtures in platinum. *Metallic materials*, 35, 1997, No.2, p. 108–119, in Czech.

[115] Wozniaková, B. and Kuchař, L.: Study of η–ε transformation of Pu–Fe and Pu–Al alloys. *Transactions of the VŠB-Technical University of Ostrava, Metallurgical Series*, XXII, 1976, No. 2, p. 207–218, in Czech.

[116] Wozniaková, B. and Kuchař, L.: Survey of the plutonium – admixture systems in solidus a liquidus portion. *Transactions of the VŠB-Technical University of Ostrava, Metallurgical Series*, XXIV, 1978, No. 2, p. 67–80, in Czech.

[117] Krištofová, D., Kuchař, L. and Wozniaková, B.: Effective distribution coefficients in plutonium. *Transactions of the VŠB-Technical University of Ostrava, Metallurgical Series*, XXVI, 1979, No. 1, p. 99–117, in Czech.

[118] Kuchař, L. and Wozniaková, B.: Periodical dependence of equilibrium distribution coefficients of admixtures in plutonium on the atomic number of admixture. *Jaderná energie*, vol. 25, 1979, No. 10, p. 382–387, in Czech.

[119] Kuchař, L. and Drápala, J. Distribution coefficients of admixtures in magnesium. *Metallic materials,* 2004, vol. 42, No. 2, p. 132–142, in Czech.

[120] Kuchař, L. jr. and Kuchař, L.: Study of solidus and liquidus curves in Re – Rh, –Os, –Ir, –Pt, –Tc binary systems in the part nearly rhenium. *Transactions of the VŠB-Technical University of Ostrava, Metallurgical Series*, XXXIV, 1988, No. 1, p. 109–124, in Czech.

[121] Kuchař, L. and Habrman, P.: Verteilung von ^{114m}In und ^{110m}Ag beim Zonenschmelzen von Antimon. *Transactions of the VŠB-Technical University of Ostrava, Metallurgical Series*, XXXIX, 1993, No. 1, p. 35–42, in German.

[122] Kuchař, L., Kuchař, L.jr. and Drápala, J.: Theory of antimony refining by zone melting. *Acta Metallurgica Slovaca*, vol. 4, Special Issue, No.3, 1998, p. 110–118.

[123] Drápala, J., Kuchař, L. and Kuchař, L.jr.: Refining of scandium by crystallization methods. *Acta Metallurgica Slovaca*, vol. 2, 1996, No. 3, p. 202–210, in Czech.

[124] Duží, P., Kuchař, L., Buhrig, E. and Drápala, J.: Contribution to correlations of equilibrium distribution coefficients of impurities in silicon. *Transactions of the VŠB-Technical University of Ostrava, Metallurgical Series*, XXXII, 1986, No. 1, p. 163–189, in Czech.

[125] Hadamovsky, H.-F.: *Werkstoffe der Halbleitertechnik*. VEB Deutscher Verlag für Grundstoffindustrie, Leipzig, 1985, 368 p., in German.

[126] Drápala, J. and Kuchař, L.: Distribution coefficients in silicon and germanium. In: *Silicon '98*, TECON Scientific, s.r.o., Rožnov p. R., 1998, p. 213–217.

[127] Drápala, J., Kuchař, L., Losertová, M., Vašek, P. und Kursa, M.: Verteilungskoeffizienten bei der Kristallisation von Halbleitern. In: *49. Berg- und Hüttenmännischer Tag*, Kolloquium 1–Kristallisation und Technik Halbleiterwerkstoffe, Technische Universität Bergakademie Freiberg, 1998, p. 84, in German.

[128] Drápala, J. and Kuchař, L.: Determining the distribution coefficient of tantalum in tungsten. *Transactions of the VŠB-Technical University of Ostrava, Metallurgical Series*, XXIV, 1978, No. 1, p. 15–27, in Czech.

[129] Drápala, J. and Kuchař, L.: Study of distribution of admixtures in tantalum. *Metallic materials*, 22, 1984, No. 6, p. 751–759, in Czech.

[130] Drápala, J. and Kuchař, L.: Distribution coefficients of admixtures in tantalum. *Transactions of the VŠB-Technical University of Ostrava, Metallurgical Series*, XXXI, 1985, No. 1, p. 1–13, in Czech.

[131] Drápala, J. und Kuchař, L.: Verteilungskoeffizienten von Beimengungen im Tantal beim Elektronenstrahl – Zonenschmelzen. *Neue Hütte*, vol. 32, 1987, No. 4, p. 154–156, in German.

[132] Kuchař, L., Drápala, J., Kuchař, L.jr., Pavlova, G.S. and Ivanov, Y.M.: Study of distribution coefficients of admixtures in tellurium. *Transactions of the*

VŠB-Technical University of Ostrava, Metallurgical Series, XXXII, 1986, No. 1, p. 97–109, in Czech.

[133] Kuchař, L., Ivanov, Y.M. and Drápala, J.: Distribution of admixtures in cadmium telluride. *Transactions of the VŠB-Technical University of Ostrava, Metallurgical Series*, XXXII, 1986, No. 1, p. 111–124, in Czech.

[134] Kuchař, L., Drápala, J. and Ivanov, Y.M.: Contribution to the refinery of semiconductor materials by means normal freezing with material removal. *Transactions of the VŠB-Technical University of Ostrava, Metallurgical Series*, XXXII, 1986, No. 1, p. 125–138, in Czech.

[135] Lejbov, V.A., Drápala, J., Ivanov, J.M. and Kuchař, L.: CdTe radiation dosimeters and technological presumptions of their production. *Transactions of the VŠB-Technical University of Ostrava, Metallurgical Series*, XXXV, 1989, No. 1, p. 69–80, in Czech.

[136] Menšík, J. and Kuchař, L.: Distributing of W, Mo and Cr in the electron beam zone melted titanium. *Transactions of the VŠB-Technical University of Ostrava, Metallurgical Series*, XXII, 1976, No. 1, p. 9–20, in Czech.

[137] Repiská, L. and Kuchař, L.: Analysis of equilibrium diagrams titanium–admixture. *Transactions of the VŠB-Technical University of Ostrava, Metallurgical Series*, XXII, 1976, No. 2, p. 1–9, in Czech.

[138] Drápala, J., Luňáček, J., Kuchař, L. and Kuchař, L.jr.: Systematic study and periodic dependence of equilibriua distribution coefficients of admixtures on the atomic number of admixtures and their importance for crystallization. *Materials Science and Engineering*, A173, 1993, p. 73–77.

[139] Luňáček, J. and Kuchař, L.: Periodical dependence of equilibrium distribution coefficients of admixtures in titanium on atomic number of admixture. *Metallic materials*, vol. 31, 1993, No. 3, p. 164–175, in Czech.

[140] Duží, P. and Kuchař, L.: Distribution of iron and aluminium in zone melted uranium. *Hutnické listy*, XXIV, 1969, No.6, p. 437–441, in Czech.

[141] Kuchař, L. and Wozniaková, B.: Study of solidus and liquidus in binary systems of uranium with carbon, manganese and vanadium. *Jaderná energie*, 16, 1970, No. 8, p. 256–261, in Czech.

[142] Wozniaková, B. and Kuchař, L.: A contribution to U–C, U–Mn, U–V binary systems. *Transactions of the VŠB-Technical University of Ostrava, Metallurgical Series*, XVI, 1970, 2, p. 79–100, in Czech.

[143] Kuchař, L. and Duží, P.: Preparation of low-alloyed uranium alloys by zone melting. *Transactions of the VŠB-Technical University of Ostrava, Metallurgical Series*, XVII, 1971, No. 3, p. 23–39, in Czech.

[144] Wozniaková, B. and Kuchař, L.: Survey of the uranium – admixture systems in solidus and liquidus portion. *Transactions of the VŠB-Technical University of Ostrava, Metallurgical Series*, XVII, 1971, No. 3, p. 41–52, in Czech.

[145] Kuchař, L., Duží, P. and Wozniaková, B.: Distribution of admixtures during zone melting of low-alloyed uranium. *Jaderná energie*, 17, 1971, No. 7, p. 217–221, in Czech.

[146] Kuchař, L.: Periodicity of the distribution coefficients of admixtures in uranium. *Nuclear energy*, 19, 1973, No. 5, p. 156–160, in Russian.

[147] Wozniaková, B. and Kuchař, L.: Transformation curves in the U–Zr, U–Cr, U–Fe, U– Mo, U–Nb, U–V binary systems. *Transactions of the VŠB-*

Technical University of Ostrava, Metallurgical Series, XIX, 1973, No. 2, p. 9–97, in Czech.

[148] Wozniaková, B. and Kuchař, L.: Computation of transformation curves in binary systems. Application on uranium–admixture systems, *Nuclear energy*, 19, 1973, No.8, p. 257–265, in Russian.

[149] Drápala, J., Kuchař, L. and Ježka, D.: Study of the uranium–admixture binary systems in portion of the solidus and liquidus curves. *Transactions of the VŠB-Technical University of Ostrava, Metallurgical Series*, XXXIII, 1987, No. 1, p. 1–17, in Czech.

[150] Drápala, J. and Kuchař, L.: Study of transformation curves in the uranium – admixture binary systems, *Transactions of the VŠB-Technical University of Ostrava, Metallurgical Series*, XXXIII, 1987, No. 1, p. 19–32, in Czech.

[151] Drápala, J. and Kuchař, L.: Survey of uranium–admixture binary systems and distribution of admixtures in uranium. *Jaderná energie*, 37, 1991, No. 8, p. 297–306, in Czech.

[152] Kuchař, L. and Drápala, J.: Die Verteilungskoeffizienten von Tantal in Wolfram beim Elektronenstrahlzonenschmelzen. *Neue Hütte*, 24, 1979, No. 5, p. 185–188, in German.

[153] Drápala, J. and Kuchař, L.: Periodical dependence of distribution coefficient of admixtures in tungsten. *Annales de chimie*, science des matériaux, 1981, No. 6, p. 611–617.

[154] Drápala, J. and Kuchař, L.: Distribution coefficients of admixtures in tungsten. *Transactions of the VŠB-Technical University of Ostrava, Metallurgical Series*, XXIX, 1983, 1, p. 1–9, in Czech.

[155] Wozniaková, B. and Kuchař, L.: The transformation curves of Zr–Al, Zr–Fe, Zr–Sn binary systems used in nuclear industry. *Transactions of the VŠB-Technical University of Ostrava, Metallurgical Series*, XVIII, 1972, No. 3, p. 49–63, in Czech.

[156] Kuchař, L. and Wozniaková, B.: Some zirconium–admixture binary systems Application on the Zr–V system. *Transactions of the VŠB-Technical University of Ostrava, Metallurgical Series*, XX, 1974, No. 3, p. 1–14, in Czech.

[157] Kuchař, L. and Wozniaková, B.: Systematics of zirconium – admixture binary systems. *Hutnické listy*, XXXI, 1976, No. 5, p. 342–347.

[158] Drápala, J. and Kuchař, L. Contribution to retrograde solubility of admixtures in binary systems. In: *Metal 99*, Tanger s.r.o., Ostrava, 1999, vol. III, p. 159–164, in Czech.

[159] Drápala, J. and Kuchař, L.: Experimental evaluation of distribution coefficients of admixtures in metals. *Metallic materials*, 38, 2000, in Czech.

[160] Kuchař, L. and Drápala, J.. *Binary systems of aluminium – admixture and their importance for metallurgy.* Ed. Delta Print Děčín, Vysoká škola báňská – TU Ostrava, Děčín, 2003, 218 p.

[161] Drápala, J., Kuchař, L., Tomášek, K. and Trojanová, Z. *Magnesium, its alloys and Mg – admixture binary systems.* ES Vysoké školy báňská – TU Ostrava, Ostrava, 2004, 172 p.

[162] Kuchař, L., Drápala, J. and Kuchař, L. jr. Distribution of admixtures in aluminium alloys and calculation of solidification interval. *Transactions of the Technical University of Košice,* 1999, no. 4, p. 11–18.

[163] Drápala, J. a Máchová, K. Equilibrium distribution coefficients of admixtures in refractory metals. *Acta Metallurgica Slovaca,* 2001, 7, no. 4, p. 441–448.

[164] Drápala, J. and Kuchař, L.. Distribution coefficients of admixtures in tin. *Acta Metallurgica Slovaca,* 2004, 10, no. 1, p. 59–66.

[165] Drápala, J. and Kuchař, L. Distribution coefficients in iron–admixture system and their significance for segregation and metallurgy. *Acta Metalurgica Slovaca*, special issue, 2004, no. 1, vol. 10, p. 291–304.

[166] Bujnošková, K., Drápala, J. and Krybus, K. The influence of technological processing on structural characteristics of tungsten and molybdenum single crystals. *Acta Metalurgica Slovaca*, special issue, 2004, no. 1, vol. 10, p. 870–873.

[167] Bujnošková, K. and Drápala, J. Study of microsegregation in molybdenum and tungsten single crystals alloyed by iridium or rhenium. *Machine Dynamics Problems*, 2004, vol. 28, no. 3, p. 69–75.

[168] Bujnošková, K. and Drápala, J. Segregation phenomena in refractory metals during zone melting. *Acta Metalurgica Slovaca*, 2005, no. 3, vol. 11, p. 301–310.

[169] Bujnošková, K. and Drápala, J. Influence of crystallization conditions on segregation characteristics of alloying elements in molybdenum and tungsten single crystals. *Gornyi informatsionno–analiticheskii bulleten'. No. 2 Funktsional'nye metallicheskie materialy.* 2007, p. 169–176.

[170] Bujnošková, K., Drápala, J. and Burkhanov, G.S. Preparation and structural characteristic of low-alloyed tungsten and molybdenum single crystals. *Acta Metallurgica Slovaca*, vol. 13, 2007, No. 1, Special issue, p. 633–639.

[171] Drápala, J., Kuchař, L., Kursa, M. and Kárník, T. Metallurgical aspects of preparation of high purity metals with defined properties. *Hutnické listy,* 1999, no. 7–8, p. 120–125, in Czech.

[172] Drápala, J. and Kuchař, L. Solidus and liquidus curves and distribution coefficients of admixtures in iron and prediction of the solidification interval in low-alloyed steels. *Hutnické listy*, LV, 2000, No. 4–7, p. 59–65, in Czech.

[173] Kuchař, L. and Drápala, J. Experimental evaluation of distribution coefficients of admixtures in metals. *Metallic materials,* 2001, vol. 39, no. 1, p. 64–76, in Czech.

[174] Drápala, J. and Bujnošková, K. Contribution to the refining metallurgy of refractory metals. *Hutnické listy*, LVII, 2002, No. 6–8, p. 62–65, in Czech.

SYMBOLS

A	–	absorbance
A_r	–	nucleon number
a	–	lattice constant, lattice parameter [nm]
a_D	–	division factor at extraction
a_i	–	activity of component i
a_i^o	–	standard activity of component i
b	–	width of (molten) zone [cm]
b_R	–	distribution number at distillation
C	–	general notation of the concentration of the element in the given phase
C^*_{maj}	–	concentration of major charge carriers [cm^{-3}]
C_{LB}, C_{SB}	–	concentration of the element B in the liquid or solid
$C_L(0)$	–	concentration of the element in the liquid phase at the boundary (x=0)
$C_L(\infty)$	–	concentration of the element in the liquid in the main volume of the melt
C_M	–	concentration of ions
$C_n(x)$	–	concentration of the element ın area x after n-th solidification process
C_0	–	average initial concentration of the element in the compound
D	–	diffusion coefficient [$\mathrm{cm}^2\ \mathrm{s}^{-1}$]
D_{LB}, D_{SB}	–	diffusion coefficient of element B in the liquid or solid [$\mathrm{cm}^2\ \mathrm{s}^{-1}$]
D_C	–	distribution ratio at extraction
E	–	intensity of the electrical field [V cm^{-1}]
E^o	–	standard electrode potential [V]
e	–	elementary charge (e =1.602 · 10^{-19}C)
F	–	Faraday constant (F = 96487 C mol^{-1})
G	–	Gibbs energy (free enthalpy) [J]
G_L, G_S	–	the molar free enthalpy of the liquid or solid phase [J mol^{-1}]
G_V	–	the molar volume free enthalpy [J mol^{-1}]
g	–	relative part of the solidified crystal
grad T_L	–	temperature gradient in the liquid [K cm^{-1}]
ΔH_D	–	the change of molar enthalpy of diffusion [J mol^{-1}]

ΔH_m^A	–	molar enthalpy of melting of substance A [J mol^{-1}]
ΔH_S	–	the change of molar enthalpy of the solution of gases [J mol^{-1}]
ΔH_{ev}	–	the change of molar enthalpy of evaporation of substance [J mol^{-1}]
I	–	electrical current [A]
I_a, I_c	–	current density (rate of electrochemical reactions of anions or cations) [A cm^{-2}]
I_p	–	limiting current densities [A cm^{-2}]
K	–	rate constant
K_d	–	rate constant of convective diffusion
K_D	–	distribution coefficient at extraction
K_F	–	distribution coefficient at distillation
K_H	–	selective coefficient at ion exchange
K_I	–	distribution coefficient at ion exchange
k	–	Boltzmann constant (k = 1.38054 · 10^{-23} J K^{-1})
k	–	distribution coefficient (general)
k_{oB}^A	–	equilibrium distribution coefficient of element B in main substance A
$k_{o\,Blim}^A$	–	limiting value of the equilibrium distribution coefficient of element B in main substance A
$k_{oE,P}^A$	–	equilibrium distribution coefficient at eutectic or peritectic temperature
k_B^{A*}	–	kinetic distribution coefficient of element B in main substance A
k_{efB}^A, k_{ef}	–	effective distribution coefficient of element B in main substance A
k_{tr}	–	equilibrium distribution coefficient of element B in main substance A in phase transformation
l	–	length [cm]
l_o	–	length of crystal (ingot) [cm]
M_A	–	relative atomic (molecular) mass of substance A
m	–	mass (weight) [kg]
m_{LB}	–	tangent to the liquidus curve
n_A	–	amount of substance A [mol]
p	–	pressure [Pa]
p^o	–	atmospheric pressure (p^o=101325 Pa)
p_i	–	partial pressure of vapour of component i above the solution [Pa]
p_i^o	–	partial pressure of vapour of component i above pure solvent [Pa]

pH	–	hydrogen factor
Q	–	amount of heat [J]
q	–	electrochemical equivalent [g C^{-1}]
q_D	–	diffusion flow, diffusion rate [mol cm^{-2} s^{-1}]
R	–	molar (universal) gas constant (R = 8.314 J mol^{-1} K^{-1})
R_H	–	Hall constant [m^3 C^{-1}]
s	–	solubility of gas
ΔS	–	change of entropy [J mol^{-1} K^{-1}]
ΔS^A_m	–	molar entropy of melting of element A [J mol^{-1} K^{-1}]
t	–	temperature [°C]
T	–	thermodynamic temperature [K]
T^A_b, t^A_b	–	boiling point of element A [K,°C]
T^A_m, t^A_m	–	melting point of element A [K,°C]
T_L,T_S	–	liquidus or solidus temperature [K]
T_E, T_P	–	temperature of eutectic or peritectic reaction [K]
ΔT_{th}	–	thermodynamic change of temperature from T^A_m [K]
ΔT	–	deviation of temperature from T^A_m [K]
U_H	–	Hall voltage [V]
V	–	volume [cm^3]
v	–	rate, macroscopic rate of solidification [cm s^{-1}]
v_A	–	rate of diffusion of compound A in the gas medium [cm s^{-1}]
w_A	–	mass fraction of substance A
W_A	–	mass fraction of substance A [wt.%]
x, x_B	–	molar fraction of component B
x_{GB}	–	molar fraction of component B in the gas phase
X, X_B	–	molar fraction of component B [mol.%]
X_{SB}, X_{LB}	–	molar fraction of component B in the solid or liquid [mol.%]
ΔZ^o_A	–	change of standard isobaric potential [J mol^{-1}]
ΔZ^o_p	–	change of standard isobaric potential of chemical reaction [J mol^{-1}]
z_a, z_k	–	valency of the anion or cation
α	–	mass transfer coefficient
α^*	–	relative change of the volume of the substance with the change of state [%]
α_D	–	distribution factor at extraction
β_R	–	distribution number at distillation
β_{Rid}	–	the relative volatility coefficient
Γ	–	ratio of residual resistivity

γ_i	– activity coefficient of component *i*
δ	– thickness of the sublaminar (diffusion) layer [cm]
ε	– absorption factor [m^2 mol^{-1}]
Θ_{cr}	– critical inclination of the boat in solidification [°]
λ	– fraction of the removed material [%]
μ_B	– chemical potential of substance B [J mol^{-1}]
μ_B^o	– standard chemical potential of substance B [J mol^{-1}]
μ_{maj}	– mobility of major charge carriers [cm^2 V^{-1} s^{-1}]
ρ_p	– resistivity [Ω cm]
ρ_0	– resistivity at temperature 0 K [Ω cm]
$\rho_{4.2K}$	– resistivity at temperature 4.2 K [Ω cm]
ρ_{295K}	– resistivity at temperature 295 K [Ω cm]
ρ_S,ρ_L	– density of solid or liquid phase [kg m^{-3}]
σ	– conductivity [S m^{-1}]
σ_a	– efficient absorption cross section of thermal neutrons [m^2]
τ	– time [s]
φ	– electrical potential [V]
Φ	– light flux [lm]

APPENDICES

The tables A2–A6 summarise the values of the distribution coefficients of admixtures in 50 main elements. The admixture elements together with their proton numbers are listed in the left column of the tables. The main elements are arranged in the first row of the tables in alphabetical order. The values of k_o, presented in the appropriate tables, are limiting values of the equilibrium distribution coefficients of the appropriate admixture elements determined by extrapolation from the solidus and liquid curves of the binary systems published in the world literature. In systems for which there is still insufficient information on the binary diagram, the values of k_o were predicted by appropriate signs "<",">" or "~". This also relates to inert gases He, Ne, Ar, Kr, Xe and Rn. In the case of lanthanides, some of the values of the distribution coefficients in the lathanide–lanthanide systems were calculated using the theory of ideal solutions.

The values of the distribution coefficients presented in these tables are a summary of a systematic study of the metal–admixture binary systems of the former Department of Non-Ferrous Metals and Nuclear Metallurgy of the VSB-TU Ostrava carried out within the framework of scientific and research activities in the last 40 years.

The values of the distribution coefficients represent the fundamental material parameter not only for refining high-purity substances by crystallisation methods, i.e., zone refining and directional solidification, where they are used for optimising the refining metallurgical process, but also for microalloying special alloys. In metallurgical practice, they can be used for an objective evaluation of the segregation characteristics (micro-inhomogeneity in dendritic or cellular growth, etc.) in primary solidification formations.

Table A1. Melting point, heat of fusion and entropy of fusion for elements

Element No.	A	T_m^A [K]	ΔH_m^A [J mol^{-1}]	ΔS_m^A [J mol^{-1}K^{-1}]
3	Li	453.6	3000	6.61
4	Be	1556	12220	7.85
5	B	2365	22600	9.56
11	Na	370.8	2597	7.00
12	Mg	923	9210	9.98
13	Al	933.45	10711	11.47
14	Si	1693	46473	27.45
19	K	336.71	2321	6.89
20	Ca	1115	8540	7.66
21	Sc	1814	16100	8.88
22	Ti	1943	17500	9.01
23	V	2183	21500	9.85
24	Cr	2136	21004	9.83
25	Mn	1519	14600	9.61
26	Fe	1811	16161	8.92
27	Co	1768	17166	9.71
28	Ni	1728	17500	10.13
29	Cu	1357.87	13263	9.77
30	Zn	692.58	7322	10.57
31	Ga	302.77	5590	18.46
32	Ge	1213	34742	28.64
33	As	883	13150	14.89
34	Se	494	6694	13.55
37	Rb	312.48	2192	7.01
38	Sr	1042	10050	9.64
39	Y	1795	17200	9.58
40	Zr	2133	20064	9.41
41	Nb	2742	26800	9.77
42	Mo	2896	27630	9.54
44	Ru	2607	24300	9.32
45	Rh	2236	21600	9.66
46	Pd	1828	16736	9.16
47	Ag	1234.93	11297	9.15
48	Cd	594.1	6192	10.42
49	In	429.63	3283	7.64
50	Sn	505	7029	13.92
51	Sb	903.75	19874	21.99
52	Te	722.57	17489	24.20

Element No.	A	T_m^A [K]	ΔH_m^A [J mol^{-1}]	ΔS_m^A [J mol^{-1}K^{-1}]
55	Cs	301.59	2096	6.95
56	Ba	977	7660	7.84
57	La	1191	11300	9.49
58	Ce	1071	9200	8.59
59	Pr	1204	10000	8.31
60	Nd	1294	10900	8.42
62	Sm	1347	11100	8.24
63	Eu	1095	10500	9.59
64	Gd	1586	15200	9.58
65	Tb	1629	16300	10.01
66	Dy	1685	17200	10.21
67	Ho	1747	17200	9.85
68	Er	1802	17200	9.54
69	Tm	1818	18400	10.12
70	Yb	1092	9200	8.42
71	Lu	1936	19300	9.97
72	Hf	2495	21767	8.72
73	Ta	3293	31400	9.54
74	W	3695	35250	9.54
75	Re	3459	34000	9.83
76	Os	3306	31800	9.62
77	Ir	2720	26200	9.63
78	Pt	2042	19700	9.65
79	Au	1337.43	12552	9.39
80	Hg	239.16	2295	9.60
81	Tl	577	4142	7.18
82	Pb	600.5	4774	7.95
83	Bi	544.37	11297	20.75
84	Po	800	12560	15.70
87	Fr	300	2100	7.00
88	Ra	973	8400	8.63
89	Ac	1323	14200	10.73
90	Th	2023	19228	9.50
91	Pa	1844.78	12341	6.69
92	U	1406	12560	8.93
93	Np	916.84	9630	10.50
94	Pu	912.5	9462	10.37

Table A2. Equilibrium distribution coefficients of admixtures in Al, Au, Cd, Ce, Co, Cr, Cu, Dy, Er

Admixture		Ag	Al	Au	Cd	Ce	Co	Cr	Cu	Dy	Er
1	H		<0.1			1.16	0.15		0.12		
2	He	<0.001	<0.001	<0.001	<0.001	<0.001	<0.001	<0.001	<0.001	<0.001	<0.001
3	Li	0.5	0.58	0.58	1.88				0.27	<1	
4	Be	0.28	0.12	0.018			0.54	0.22	0.33		
5	B		0.16		0.12	<1	0.01	0.073	0.06	<1	<1
6	C					0.6	0.28	0.011	~0.1		
7	N		<0.1			1.57		0.17			
8	O	0.09			<1		<0.1	~0.03	0.019		
9	F										
10	Ne	<0.001	<0.001	<0.001	<0.001	<0.001	<0.001	<0.001	<0.001	<0.001	<0.001
11	Na		0.017	~0.01	0.09						
12	Mg	0.61	0.31	0.56	1.40	0.67	~0.1		0.20	0.47	0.71
13	Al	0.52	1	0.36	<0.1	<0.1	0.67	0.80	0.81		0.07
14	Si	<0.1	0.094	0.035		0.31	0.47	0.44	0.42		
15	P	0.01	<0.1	<0.1	<0.1		0.02	~0.09	0.091		
16	S	0.017	<0.01	<0.1	<0.1		~0.01	~0.03	0.024	<1	
17	Cl										
18	Ar	<0.001	<0.001	<0.001	<0.001	<0.001	<0.001	<0.001	<0.001	<0.001	<0.001
16	K										
19	Ca	<0.1	0.06	0.088	<0.1	<1			0.05		
21	Sc	0.39	0.23	0.70		1.70	0.03	~0.05	0.027		<1
22	Ti	0.97	9.1	1.82	<0.1	0.30	0.43	0.42	0.31		0.18
23	V	~1	3.71	1.70		1.83	0.76	0.63	2.46		0.28
24	Cr	~1	1.7	1.56	<0.1	0.39	0.87	1	0.54		<1
25	Mn	1.06	0.76	0.59	0.17	0.24	0.65	0.58	0.45	<1	<1
26	Fe	0.1	0.02	0.87	<0.1	0.04	0.96	0.66	1.32	<1	<1
27	Co	1.89	0.02	0.72	<0.1	<0.1	1	0.56	1.44	<1	<1
28	Ni	0.58	0.0068	0.70	0.04	<0.1	0.97	0.41	1.48	<1	<1
29	Cu	0.56	0.14	0.63	0.09	0.07	0.62	~0.5	1	0.08	<1
30	Zn	0.49	0.43	0.36	0.39	<0.1	0.55		0.72	<1	<1
31	Ga	0.41	0.086	0.11	0.25	<0.1	0.40	0.87	0.66	0.33	<1
32	Ge	0.19	0.08	0.063	0.01	<0.1	0.50	0.41	0.33	<1	<1
33	As	0.19	<0.1	0.026	<0.1		0.14		0.16	<1	<1
34	Se	0.20	<0.01	<0.1	<0.1					<1	<1
36	Kr	<0.001	<0.001	<0.001	<0.001	<0.001	<0.001	<0.001	<0.001	<0.001	<0.001
38	Sr	<0.1	0.045	~0.01	<0.1		<0.1		~0.025		
39	Y	0.08	0.027		<0.1	2.01	0.02	0.23	0.0072	1.05	0.97
40	Zr	0.051	1.97	1.01			0.06	0.077	0.019	>1	0.44
41	Nb		4.59	2.24			0.23	0.33	0.18		
42	Mo	0.31	1.56	>1			0.84	0.83	<1		
43	Tc		<1	>1		<1		1.15	>1		

Admixture		Ag	Al	Au	Cd	Ce	Co	Cr	Cu	Dy	Er
44	Ru	>1	<1	1.06		0.58	0.64	0.69	>1	<1	
45	Rh	>1	<1	1.68			0.82	0.44	2.15	<1	0.06
46	Pd	1.95	<1	2.29	0.03		0.45	0.27	1.09	<1	
47	Ag	1	0.32	0.96	1.86	0.17		0.18	0.34	<1	
48	Cd	0.73	0.062	0.50	1				0.27	<1	
49	In	0.56	0.012	0.48	0.20	0.44	<0.1		0.45	0.36	0.38
50	Sn	0.46	0.01	0.17	0.11	1.2	0.27	0.08	0.18	<1	0.08
51	Sb	0.28	0.034	0.038	0.02		0.15	0.12	0.088	<1	<1
52	Te	0.11	<0.01	<0.01		<0.1	<0.1		0.011	<1	<1
54	Xe	<0.001	<0.001	<0.001	<0.001	<0.001	<0.001	<0.001	<0.001	<0.001	<0.001
56	Ba				<0.1	0.25					
57	La	0.0062	0	0.009	<0.1	1.13	<0.1	0.21	~0.06		
58	Ce	0.01	0.0032	0.007	<0.1	1	0.1	0.1	0.027	0.69	0.65
59	Pr	0.0019	0		<0.1	1.15		0.19	~0.08	0.75	0.71
60	Nd	0.022	0	0.012	<0.1			0.30	~0.07	0.85	0.44
61	Pm				<0.1					0.78	0.73
62	Sm	0.04	<0.1	0.017	<0.1	1.19	0.09	~0.18	0.16	0.82	0.78
63	Eu	0			<0.1	0.84				0.67	0.63
64	Gd	0.067	<0.1	0.049	<0.1	1.40	0.074	~0.19		0.93	0.87
65	Tb	0.10	0.016	0.06	<0.1			0.17		0.96	0.89
66	Dy	0.1	<0.1	0.15	<0.1		<0.1			1	0.92
67	Ho	0.13	<0.1	0.26	<0.1			0.19	0.003	1.05	0.96
68	Er	0.18	<0.1	0.19	<0.1		0.02			1.08	1
69	Tm	0.21	0.0075	0.25	<0.1	1.80		0.40		1.11	1.02
70	Yb	0.11	<0.1	0.24	<0.1				0.002	0.7	0.66
71	Lu	0.27		0.33	<0.1			0.22		1.2	1.1
72	Hf	<1	2.92	<1		<1	0.1	0.063	0.053		<1
73	Ta		6.07	0.57			0.25	0.43	>2	<1	<1
74	W		>1				1.09	1.41	>2		
75	Re		<1				1.22	1.45	>1		
76	Os	>1					1.45	1.20	>1		
77	Ir	>1				<0.1	0.80	0.52	2.86		
78	Pt	2.41	0.39	1.98	0.03		0.64	0.25	2.03		
79	Au	1.17	0.035	1	0.54	<0.1	0.18	0.089	0.64	<1	<1
80	Hg	0.51	0.17	0.22	0.35				0.093		
81	Tl	0.28	<0.1	0.15	<0.1	0.33			0.024		
82	Pb	0.27	0.056	0.013	0.036	<0.1			0.023	0.39	<1
83	Bi	0.2	0.033	0.011	0.002	0.31	<0.1		0		
86	Rn	<0.001	<0.001	<0.001	<0.001	<0.001	<0.001	<0.001	<0.001	<0.001	<0.001
90	Th	0.016	0.05	0.0012		1.47	0.24		0.0007	1.72	1.21
92	U	0.31	0.0024	0.025	<0.1		<0.1	0.08	~0.025		
93	Np				<0.1						
94	Pu		0.0035		<0.1	0.74	<0.1	~0.05	0.018		

Table A3. Equilibrium distribution coefficients of admixtures in Fe, Ga, Gd, Ge, Hf, Ho, In, Ir, La, Lu

Admixture		Fe	Ga	Gd	Ge	Hf	Ho	In	Ir	La	Mg
1	H	~0.3		>1	$<10^{-7}$					1.15	0.92
2	He	<0.001	<0.001	<0.001	$<10^{-7}$	<0.001	<0.001	<0.001	<0.001	<0.001	<0.001
3	Li				~0.003			~0.06		>1	0.68
4	Be	0.64			~0.081	<0.01					
5	B	0.05		<0.1	~17	0.15	<1				
6	C	0.19		~0.1		0.32			~0.2	0.36	
7	N			>1						>1	
8	O	0.15		>1	~0.001	2.59				>1	
9	F				$<10^{-6}$						
10	Ne	<0.001	<0.001	<0.001	$<10^{-7}$	<0.001	<0.001	<0.001	<0.001	<0.001	<0.001
11	Na		<1		$<10^{-5}$			1.24		>1	
12	Mg	0.62		0.48	~0.0001		0.38	1.21	~0.1	0.60	1
13	Al	0.88	<1	0.04	0.01	0.60	<1	<1	~0.5	<0.1	0.29
14	Si	0.66	<0.1	0.49	5.5	0.17		<1			0.0015
15	P	0.14			0.1			<1			
16	S	0.06		<0.1	~0.001			<1		<0.1	
17	Cl				$<10^{-6}$					<0.1	
18	Ar	<0.001	<0.001	<0.001	$<10^{-7}$	<0.001	<0.001	<0.001	<0.001	<0.001	<0.001
16	K		<1		$<10^{-6}$						
19	Ca		0.95		$<10^{-6}$					0.63	0.046
21	Sc	0.06		0.86		<1			~0.1	0.88	2.18
22	Ti	0.35		0.52		0.63			0.39	<1	1.27
23	V	0.86	0.46	<1	$\sim 3.10^{-7}$	0.43	0.05		<1	0.24	
24	Cr	0.90		0.26		0.46	0.35		0.76	<0.1	
25	Mn	0.69		<1	$<10^{-8}$	0.12	<1	<1	<1	0.04	1.37
26	Fe	1	0.35	0.05	$<10^{-5}$	0.14	<1	<1	0.46	0.03	
27	Co	0.84	<1	<0.1	$<10^{-6}$	0.33	<1	<1	0.33	<0.1	
28	Ni	0.69	<1	0.06	0	0.04		<1	<1	<0.1	0.0021
29	Cu	0.78	<0.1	0.33	0	0.27		~0.06	0.33	<0.1	0.0007
30	Zn	0.77	0.15	<1	0			0.46		<0.1	0.24
31	Ga	0.81	1	0.47	0.04	0.18		0.27		0.32	0.082
32	Ge	0.79	<0.1	<0.1	1	0.21	<1	<1		<0.1	0.055
33	As	0.21			0.02			<1			
34	Se				~0.001						
36	Kr	<0.001	<0.001	<0.001	$<10^{-7}$	<0.001	<0.001	<0.001	<0.001	<0.001	<0.001
38	Sr				$<10^{-6}$						0.0045
39	Y	0.06		1.27		0.03	1.03		<0.1	1.61	0.19
40	Zr	0.28		0.13		0.80			0.42		3.77
41	Nb	0.37		0.82		0.78			0.87	1.27	
42	Mo	0.73		0.83		0.48			0.93	>1	
43	Tc	1.03							<1		

Admixture		Fe	Ga	Gd	Ge	Hf	Ho	In	Ir	La	Mg
44	Ru	1.04		<1					0.92		
45	Rh	0.68							0.82	<1	
46	Pd	0.47		0.36		0.25	<1		~0.68		0.023
47	Ag	0.042	0.45	<1	$\sim 3.10^{-7}$		<1	~0.09	0.32	0.22	0.10
48	Cd		~0.1	0.45	$<10^{-5}$		<1	0.78		0.43	0.43
49	In	0.21	0.03	0.60	0		0.48	**1**		0.41	0.43
50	Sn	0.28	0		0.02	0.44	<1	0.69		<0.1	0.26
51	Sb	0.15			0		<1	<0.1	~0.01	<0.1	
52	Te				0		<1	2.82		<0.1	
54	Xe	<0.001	<0.001	<0.001	$\sim 3.10^{-7}$	<0.001	<0.001	<0.001	<0.001	<0.001	<0.001
56	Ba			0.36	$<10^{-6}$					<0.1	0.0013
57	La	0.05		<1	0.04				~0.08	**1**	0.025
58	Ce	0.02		0.67	0.03	0.53	0.67		~0.09	0.9	0.025
59	Pr			<1	0.05		0.73			1.04	0.027
60	Nd	0.08		0.80	0.02		0.77			1.1	0.14
61	Pm			<1			0.75			1.63	
62	Sm	0.01		0.86			0.8			1.52	0.22
63	Eu			<1			0.65			<1	
64	Gd	0.07		**1**			0.87			1.61	0.21
65	Tb	0.06		1.04			0.92			>1	0.27
66	Dy			1.06			0.95				0.35
67	Ho	<0.1		1.11			**1**				0.38
68	Er	0.1		1.14		~0.1	1.02			<0.1	0.34
69	Tm	<0.1					1.06				0.51
70	Yb			0.85			0.68		~0.1		0.061
71	Lu	<0.1					1.14				0.59
72	Hf	0.22				**1**			0.43		
73	Ta	0.44		<1	~0.008	0.76			0.78	<0.1	
74	W	0.83				0.44			0.65		
75	Re	1.10				0.38			1.16		
76	Os	1.41				0.24			1.48		
77	Ir	1.11				0.17			**1**	<0.1	0.026
78	Pt	0.72	<0.1	<1	$\sim 5.10^{-7}$	~0.1		0.26	0.76		
79	Au	0.36			$7.8.10^{-6}$	0.25	<1	<0.1	~0.25	<0.1	0.12
80	Hg		0.08		~0.0001			0.59			0.037
81	Tl		<0.1	0.31	0			1.06		0.28	0.29
82	Pb		<0.1	0.24	~0.0017			1.044		<0.1	0.24
83	Bi		<0.1	<0.1	0			0.33	~0.01	0.28	0.081
86	Rn	<0.001	<0.001	<0.001	$<10^{-7}$	<0.001	<0.001	<0.001	<0.001	<0.001	<0.001
90	Th	0.06		2.20		0.28			<0.1	1.20	0.12
92	U	<0.1				0.55			0.05		
93	Np										
94	Pu	0.21				~0.5				0.71	0.14

Table A4. Equilibrium distribution coefficients of admixtures in Mo, Nb, Nd, Ni, Os, Pb, Pd, Pr, Pt, Pu

Admixture		Mo	Nb	Nd	Ni	Os	Pb	Pd	Pr	Pt	Pu
1	H			>1	0.37				1.06		
2	He	<0.001	<0.001	<0.001	<0.001	<0.001	<0.001	<0.001	<0.001	<0.001	<0.001
3	Li				0.22		0.24	0.35			
4	Be	0.01	0.25		0.50			0.018		0.046	<1
5	B	0.038	0.034		0.018	~0.1		0.33		~0.1	
6	C	0.038	0.41		0.19	0.30		0.68		~0.3	<1
7	N	0.019	0.77								
8	O		0.20		0.062						<1
9	F										
10	Ne	<0.001	<0.001	<0.001	<0.001	<0.001	<0.001	<0.001	<0.001	<0.001	<0.001
11	Na		~0.01				0.72				
12	Mg		0.053		0.012		0.34	0.58	0.63		0.2
13	Al	0.42	0.38		0.85	~0.3		0.50	<1	0.71	3.85
14	Si	0.080	0.14	<1	0.41			0.03		0.06	0.1
15	P				0.018			0.0019	0.56	<0.1	
16	S	0.17			0			0.031			
17	Cl										
18	Ar	<0.001	<0.001	<0.001	<0.001	<0.001	<0.001	<0.001	<0.001	<0.001	<0.001
16	K										
19	Ca			0.65	0.0052		0.67	~0.02	<1	<0.1	
21	Sc	0.022	<0.1	1.16	0.058	~0.1		0.71	1.72	~0.1	3.06
22	Ti	0.60	0.57	0.32	0.49	0.39		0.45		1.11	2.05
23	V	0.73	0.60		0.94	0.69		0.49		1.03	
24	Cr	0.57	0.44	0.77	0.83	<1		0.67	0.55	1.09	
25	Mn	~0.2			0.58	<1	0.43	0.79		0.71	0.13
26	Fe	0.27	0.13	0.05	0.91	0.50		0.62		0.69	0.07
27	Co	0.22	0.023		1.03	0.44		0.58		0.69	0.04
28	Ni	0.042	0.068		1	~0.1	0.058	0.55	<1	0.54	0.16
29	Cu		0.068		0.79		0.083	0.66	<1	0.88	0.70
30	Zn			0.23	0.62		0.14	0.79	0.26	<1	
31	Ga	0.17	0.34	0.04	0.49		0.14	0.36	<1	<1	2.34
32	Ge	0.063	0.18	0.07	0.46			0.048	<1	~0.1	
33	As			0.18	0.22		0.023		<1	<0.1	
34	Se							0.011			
36	Kr	<0.001	<0.001	<0.001	<0.001	<0.001	<0.001	<0.001	<0.001	<0.001	<0.001
38	Sr			0.55	<0.1			~0.1		<0.1	
39	Y	~0.05		1.49	0.013	~0.1		0.26	1.76	~0.1	
40	Zr	0.14	0.41		0.05	0.54		1.17		2.18	4.37
41	Nb	0.87	1		0.51	0.43		1.31		1.47	
42	Mo	1	1.095		0.84	0.76		1.50		1.39	
43	Tc	0.37			1.11	<1		1.41		>1	

Admixture		Mo	Nb	Nd	Ni	Os	Pb	Pd	Pr	Pt	Pu
44	Ru	0.34	0.39		1.21	0.86		1.17		1.47	0.25
45	Rh	0.27	0.32		1.22	~0.5	0.28	1.26		1.38	0.18
46	Pd	0.17	0.54		0.51	0.11		1		0.88	0.53
47	Ag			0.15	0.39	<0.1	0.064	0.67	0.12	0.53	
48	Cd				<0.1		0.32	~0.7		<1	
49	In			0.50	0.20		0.74	0.68	<1	<1	
50	Sn		0.39	0.30	0.28		0.62	0.57	<1	0.39	
51	Sb		0.26	0.29	0.23		0.33	0.47	0.18	0.21	
52	Te			<1				0.12	<1	<0.1	
54	Xe	<0.001	<0.001	<0.001	<0.001	<0.001	<0.001	<0.001	<0.001	<0.001	<0.001
56	Ba			0.6	0.082		0.11		<1		
57	La	~0.07	0.054	0.91	<0.1	<0.1	0.21	~0.1		~0.1	
58	Ce		0.31	0.84	0.024			0.18	0.91		0.90
59	Pr			0.96	<0.1	<0.1		0.31	1	<0.1	
60	Nd			1				0.10	1.09		0.55
61	Pm			1.05					1.11		
62	Sm		0.13	1.02			0.17	0.16	1.12		0.83
63	Eu			0.84	0.027			0.26	0.9	<0.1	
64	Gd	0.088		1.03	0.093		0.18	0.29	1.29	0.05	
65	Tb			1.36				0.27	1.53		
66	Dy			1.46				0.38	1.63		
67	Ho			1.51				0.30	1.7		
68	Er			1.2	<0.1			0.30	1.77		
69	Tm										
70	Yb			0.89				0.52		<0.1	
71	Lu	<0.1									
72	Hf	0.47	0.54		0.052	0.46		1.22		>1	5.49
73	Ta	1.32	1.28		0.62	0.31		1.91		2.15	
74	W	1.54	1.22		1.64	0.85		1.60		1.73	
75	Re	0.85	0.96		2.20	~1.05		1.78		2.02	
76	Os	0.79	0.62		1.10	1		3.46	0.06	>1	0.19
77	Ir	0.24	0.39		1.96	0.71		1.92	0.03	1.11	
78	Pt	0.23	0.44	<1	0.94	~0.5		1.30	0.07	1	0.26
79	Au	0.017	0.41		0.26		0.011	0.91	0.12	0.71	
80	Hg				~0.1		0.67		0.28	<1	
81	Tl			0.29	0.30		1.17	0.084	0.29	<1	
82	Pb				0.13		1	0.41	0.28	~0.1	
83	Bi			0.1	<0.1		0.78	0.33	0.21	<0.1	
86	Rn	<0.001	<0.001	<0.001	<0.001	<0.001	<0.001	<0.001	<0.001	<0.001	<0.001
90	Th	0.046		4.44	0	<0.1		~0.01	1.39	~0.1	0.60
92	U	0.19	0.53		0.032	0.04		0.54		0.19	0.54
93	Np	0.42									0.70
94	Pu	~0.15		0.80	0.11	<0.1		0.51	0.83	~0.1	1

Table A5. Equilibrium distribution coefficients of admixtures in Re, Rh, Ru, Sb, Sc, Si, Sm, Sn, Ta, Tb

Admixture		Re	Rh	Ru	Sb	Sc	Si	Sm	Sn	Ta	Tb
1	H							>1			>1
2	He	<0.001	<0.001	<0.001	<0.001	<0.001	$<10^{-7}$	<0.001	<0.001	<0.001	<0.001
3	Li				0.29		0.01	<1			<1
4	Be			0.05		<1	~0.1	<1			<1
5	B	0.014	~0.09	0.045		<1	0.27	<1		0.048	<1
6	C	0.19	0.14	0.16		2.34	~0.07			0.52	
7	N						$<10^{-7}$			0.69	
8	O					1.87	<1			0.10	>1
9	F						$<10^{-6}$				<0.1
10	Ne	<0.001	<0.001	<0.001	<0.001	<0.001	$<10^{-7}$	<0.001	<0.001	<0.001	<0.001
11	Na				<0.05			<1			<1
12	Mg				0.015	0.63	$\sim 8.10^{-6}$	0.70			0.49
13	Al	0.037	~0.2	0.14		0.13	0.02	<1	0.43	0.081	<1
14	Si	0.065	~0.08		0.011		1	<1		0.11	<1
15	P		~0.01	~0.03	0.026		0.35	<1			
16	S				0.054		$\sim 10^{-5}$	<0.1			
17	Cl				<0.05		$<10^{-6}$	<0.1			<0.1
18	Ar	<0.001	<0.001	<0.001	<0.001	<0.001	$<10^{-7}$	<0.001	<0.001	<0.001	<0.001
16	K				<0.05			<1			<1
19	Ca				0.32	0.35		<1			<1
21	Sc	0.13	~0.1	0.09		1		1.22			
22	Ti	0.085	0.57	0.33	<1	0.51	$\sim 2.10^{-6}$		0.063	0.61	<1
23	V	0.15	0.71	<1		0.20	$\sim 4.10^{-6}$	<1		0.47	
24	Cr	0.21	0.66	0.66		0.22	$\sim 8.10^{-6}$	<1		0.32	0.33
25	Mn		<1	<1		<1	$\sim 8.10^{-6}$	<1			<1
26	Fe	~0.5	0.63	0.93		0.1	0	<1		0.09	0.34
27	Co	0.32	0.33	0.53	<0.05	0.02	0	<1		0.22	
28	Ni	0.22	0.68	0.63	0.024	<1	0	<1	0.013	0.36	<1
29	Cu	~0.4	0.48	~0.2	0.045		0	<1	0.006		<1
30	Zn		~0.25				0	0.34	0.032		
31	Ga		~0.1		0.16	<1	0.01	0.30	0.17		0.25
32	Ge		~0.05	<0.1	0.37	0.11	0.39	<1			<1
33	As				0.61		0.24	<1			<1
34	Se		~0.01		<0.5		$\sim 2.10^{-5}$	<0.1			<1
36	Kr	<0.001	<0.001	<0.001	<0.001	<0.001	$<10^{-7}$	<0.001	<0.001	<0.001	<0.001
38	Sr				0.17	0.39					<1
39	Y	<0.1	~0.1	~0.05		0.63		1.36			>1
40	Zr	0.026	0.46	0.12		1.14	$<10^{-8}$	1.95		0.15	
41	Nb	0.12	0.98	0.57		0.17				0.80	
42	Mo	0.19	1.5	0.82		0.18	$\sim 6.10^{-8}$			0.83	
43	Tc	0.49	>1	<1		<1					

Admixture		Re	Rh	Ru	Sb	Sc	Si	Sm	Sn	Ta	Tb
44	Ru	0.61	1.12	1		0.02		0.03		0.48	<1
45	Rh	0.38	1	0.84	<1			<1		0.31	<1
46	Pd	0.057	0.76	0.34	<1			0.07		0.44	<1
47	Ag		0.72	~0.2	0.28	<1	0	<1	0.034		0.89
48	Cd		~0.6		0		$\sim10^{-5}$	<1	0.21		<1
49	In		~0.5			0.53	$\sim4.10^{-4}$	0.42	0.39		0.73
50	Sn		0.27		0.37		0.01	0.03	1		<1
51	Sb		0.18		1		0.02	<1	1.95		0.46
52	Te		~0.01		0.07		$\sim8.10^{-6}$	<1			<1
54	Xe	<0.001	<0.001	<0.001	<0.001	<0.001	$<10^{-7}$	<0.001	<0.001	<0.001	<0.001
56	Ba				<0.1	<1		0.23			0.29
57	La		~0.09	~0.02		0.21		0.88			0.63
58	Ce			0.05		0.54		0.86			
59	Pr					<1					
60	Nd				0.30	0.56		0.95			
61	Pm										
62	Sm							1			
63	Eu							0.8			0.69
64	Gd	0.021		<0.1		0.65		1.20			0.97
65	Tb	0.022						1.29			1
66	Dy			<0.1				1.36			1.03
67	Ho							1.69			1.08
68	Er	0.025	~0.1			0.74		1.47			1.12
69	Tm										
70	Yb		~0.1	~0.1							
71	Lu										
72	Hf	0.011	~0.8							0.42	
73	Ta	0.13	1.18	0.6		<1	$\sim10^{-7}$			1	
74	W	0.34	1.70	0.88		<1				0.98	<1
75	Re	1	5.22	1.48		<1		<1		0.58	0.06
76	Os	0.87	~2.5	1.45				<1		0.77	<1
77	Ir	0.64	1.53	1.02				<1		0.24	<1
78	Pt	0.40	0.98	0.62				<1		0.26	<1
79	Au		0.26	~0.2		<1	0	0.02	0.036	0.23	<1
80	Hg		~0.1					<1	0.20		<1
81	Tl						$\sim10^{-4}$	<1	0.058		<1
82	Pb				0.15		$\sim5.10^{-4}$	0.55	0.058		<1
83	Bi		~0.01	~0.01	0.24		0	<0.1	0.27		
86	Rn	<0.001	<0.001	<0.001	<0.001	<0.001	$<10^{-7}$	<0.001	<0.001	<0.001	<0.001
90	Th		0.04	~0.08		0.74				0.013	1.32
92	U	0.058	0.03	~0.1		0.54	0.14			0.020	
93	Np										
94	Pu		<0.1	~0.1		0.73		0.87			

Table A6. Equilibrium distribution coefficients of admixtures in Te, Th, Ti, Tm, U, V, W, Y, Yb, Zr

Admixture		Te	Th	Ti	Tm	U	V	W	Y	Yb	Zr
1	H			0.58		0.60	0.40				
2	He	<0.001	<0.001	<0.001	<0.001	<0.001	<0.001	<0.001	<0.001	<0.001	<0.001
3	Li	0.058				<0.1			<1		
4	Be		0.026	0.12		0.34	0.38	0.19	<1		0.12
5	B			0.081	<1	<0.1	0.07	0.023		<1	0.13
6	C		0.25	0.37		0.20	0.40	0.018	1.33		0.30
7	N		1.71	6.65		<0.1	0.76		>1		1.22
8	O			1.79		0.09	0.73		1.48		1.85
9	F								<0.1		
10	Ne	<0.001	<0.001	<0.001	<0.001	<0.001	<0.001	<0.001	<0.001	<0.001	<0.001
11	Na	<0.1							<1		
12	Mg	<0.1	<0.1		0.44	<0.1			0.55	0.35	
13	Al	0.0035	0.11	1.41		0.57	0.94	0.061	<1	0.41	0.38
14	Si		<0.1	0.44		0.14	0.59	0.044	<1		0.05
15	P		<0.1	0.023					<1		
16	S	0.30		0.0003		<0.1		<0.1	<1		
17	Cl								<0.1		
18	Ar	<0.001	<0.001	<0.001	<0.001	<0.001	<0.001	<0.001	<0.001	<0.001	<0.001
16	K								<1		
19	Ca	0.01							<1	0.95	
21	Sc		0.52	0.38			0.21		0.64	~1	
22	Ti		<0.1	1		1.39	0.61	0.28	0.25		0.70
23	V		0.018	0.85		0.56	1	0.23	0.21	<1	0.34
24	Cr	<1	<0.01	0.59		0.17	0.75	0.49	<1		0.18
25	Mn	<1	0.03	0.37		0.17	0.69		<1	<1	0.21
26	Fe	<1	<0.1	0.26		0.1	0.56	0.035	0.1	<1	0.12
27	Co	<1	<0.1	0.21		0.05	0.63	0.013	<1	<1	0.07
28	Ni	<1	<0.1	0.21		0.03	0.33	0.010	0.03	<1	0.07
29	Cu	0.01	<0.1	0.20		<0.1	0.47		<1	<1	0.11
30	Zn	0.03	<0.1			<0.1			<1	<1	0.29
31	Ga	<1		0.78		<0.1	0.85		<1	<1	0.17
32	Ge	0.3	<0.1	0.54		0.26	0.33	<0.1	0.1	<1	0.19
33	As	0	<0.1	0.061					<1		
34	Se	0.45	<0.1						<1		
36	Kr	<0.001	<0.001	<0.001	<0.001	<0.001	<0.001	<0.001	<0.001	<0.001	<0.001
38	Sr								0.31		
39	Y		0.83	0.15	~1			<0.1	1	~1	0.12
40	Zr	<1	0.44	0.76		1.92	0.10	0.069	0.34		1
41	Nb		0.10	1.21		2.27	0.81	0.41	0.04		0.65
42	Mo	<1	0.25	1.69		1.34	1.39	0.6	<0.1		0.67
43	Tc			1.51			1.20		<1		<1

Admixture		Te	Th	Ti	Tm	U	V	W	Y	Yb	Zr
44	Ru		0.02	0.54		0.23	1.03	0.45	<1	<1	0.19
45	Rh		0.02	0.40		0.20	0.72	0.093	<1	<1	0.13
46	Pd		0.010	0.60		0.32	0.64	0.046	<1	<1	0.22
47	Ag	~0.007		0.69			0.46		<1		0.30
48	Cd	~0.90		0.39		<0.1			<1	0.33	<1
49	In		0.17		0.36		0.79		0.41	0.29	
50	Sn	~0.65		0.90		<0.1	0.84		0.11	<1	0.27
51	Sb	~0.48		0.40		<0.1			0.28	<1	
52	Te	**1**		0.19					<1	<1	0.12
54	Xe	<0.001	<0.001	<0.001	<0.001	<0.001	<0.001	<0.001	<0.001	<0.001	<0.001
56	Ba								<1	0.75	
57	La		0.68	0.35			0.015		0.68		
58	Ce	<1	0.48	0.25	0.65	0.13	0.066		0.72	0.98	0.41
59	Pr		0.53	0.094	0.71				<1	1.1	
60	Nd		0.65	0.18	0.75				0.7	0.86	
61	Pm				0.73				<1	1.26	
62	Sm				0.77				0.73	1.25	0.31
63	Eu				0.63				<1	0.99	
64	Gd		0.77	0.18	0.86		0.072		0.83	1.18	0.13
65	Tb		0.76		0.88				0.84	1.8	
66	Dy		0.81	0.14	0.91		0.061		0.86	1.94	0.47
67	Ho	<1	0.94		0.95		0.052		0.97	2.02	
68	Er		0.84	0.18	0.99		0.046		0.98	2.09	0.53
69	Tm			0.11	**1**				<1	2.24	
70	Yb	<1			0.67				<1	**1**	
71	Lu		0.84		1.08				<1	2.52	
72	Hf		0.58	0.92		1.98	0.18	0.13	0.28	<1	1.13
73	Ta		0.03	2.13		2.83	0.90	0.79	0.01		0.82
74	W		<0.1	1.98		1.28	1.76	**1**	<1		0.40
75	Re		0.03	1.61		0.83	2.07	0.43	<1		0.27
76	Os		<0.1	0.74		0.50	2.54	0.44	<1		0.21
77	Ir		<0.1	0.57		0.24	1.06	0.15	<1	<1	0.09
78	Pt		0.02	0.32		0.28	0.75	0.090	<1	<1	0.12
79	Au	0.002	0.05	0.37		0.55	0.49		<1	<1	0.14
80	Hg	0.05				<0.1			<1		
81	Tl	<1							<1		
82	Pb	0.13	0.06	0.39		<0.1			0.38	0.47	~0.18
83	Bi	0.0002	<0.1	0.39			0.054		<1		0.26
86	Rn	<0.001	<0.001	<0.001	<0.001	<0.001	<0.001	<0.001	<0.001	<0.001	<0.001
90	Th		**1**			0.25	0.019	<0.1	1.45		0.55
92	U		0.24	0.80		**1**	0.077	0.001	<1		0.64
93	Np					0.68			<1		
94	Pu		0.43	0.28		0.54	0.34		0.25		0.50

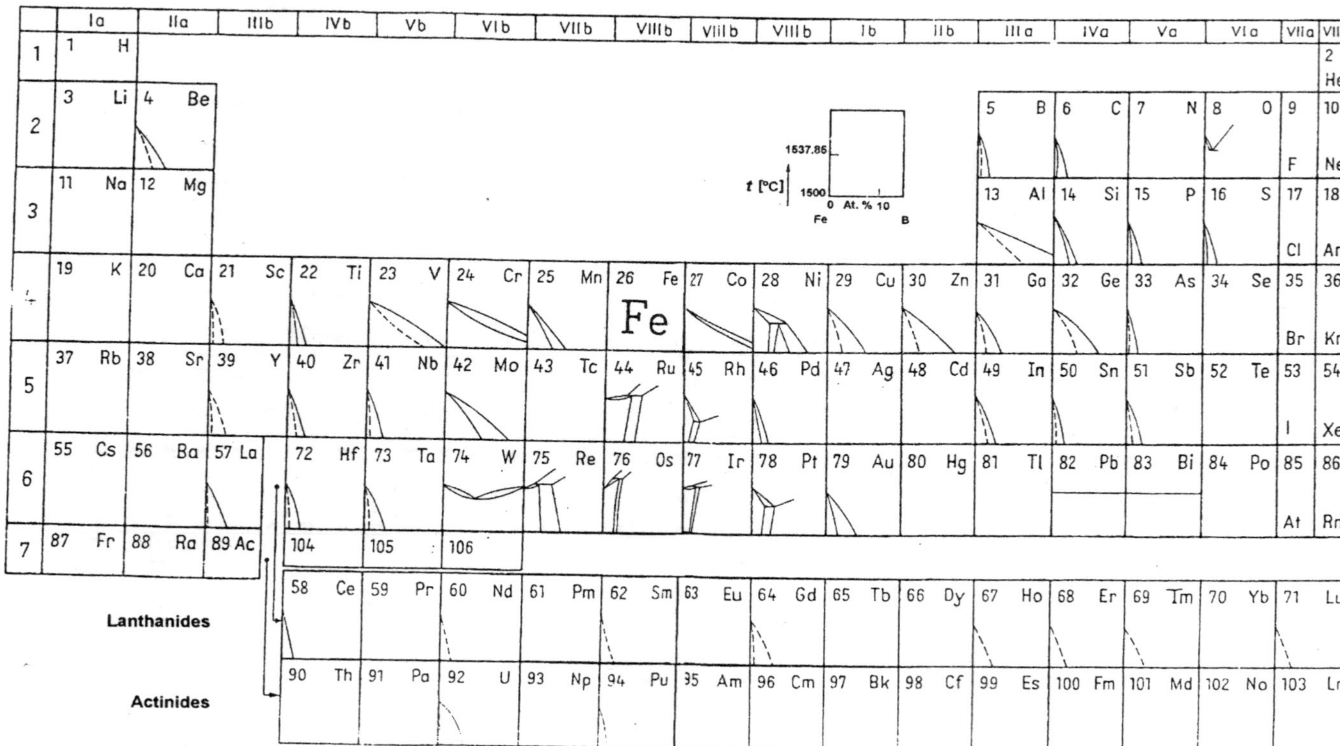

Fig. A2. Types of iron–admixture binary systems in the liquids and solidus regions.

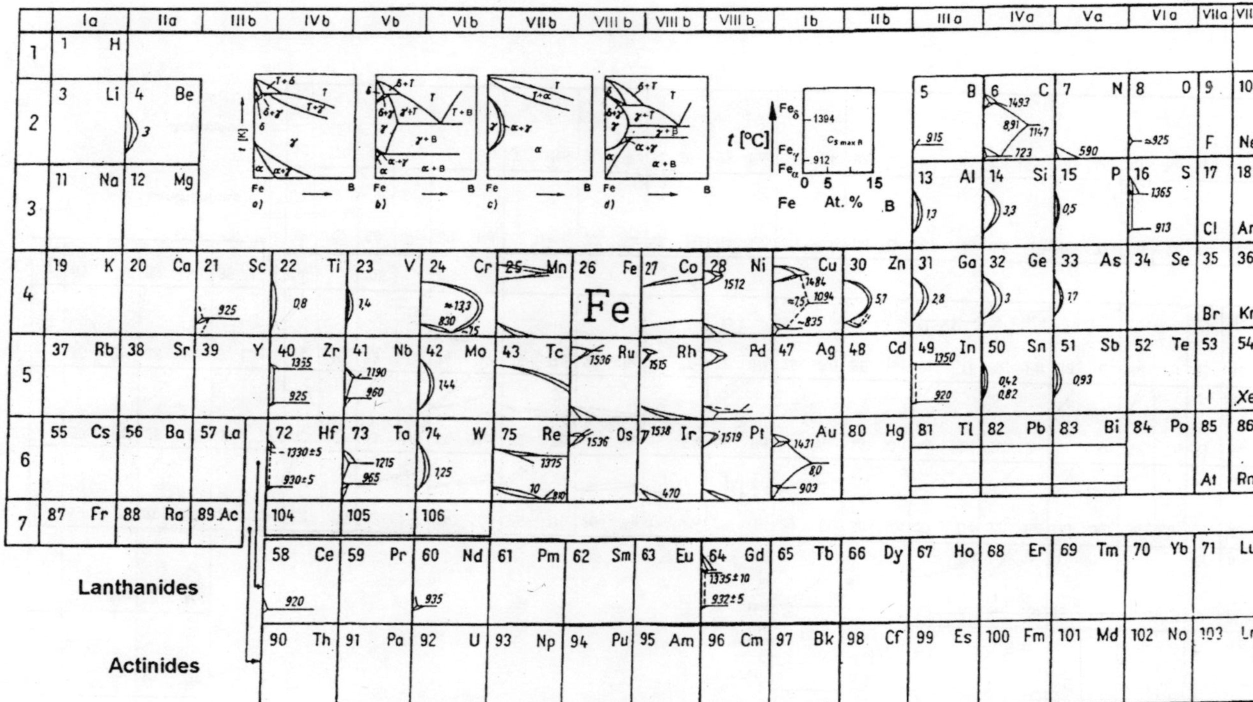

Fig. A3. Types of iron–admixture binary systems in the region of δ–γ and γ–α transformation curves.

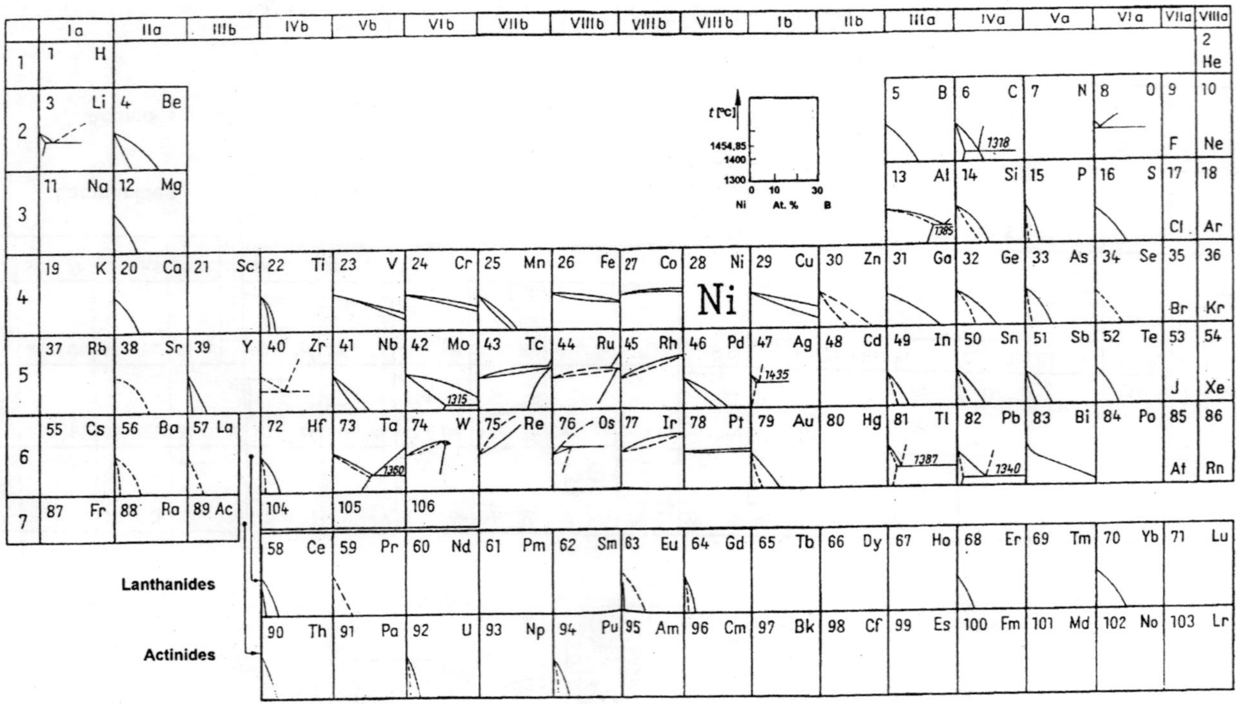

Fig. A4. Types of nickel–admixture binary systems in the liquids and solidus regions.

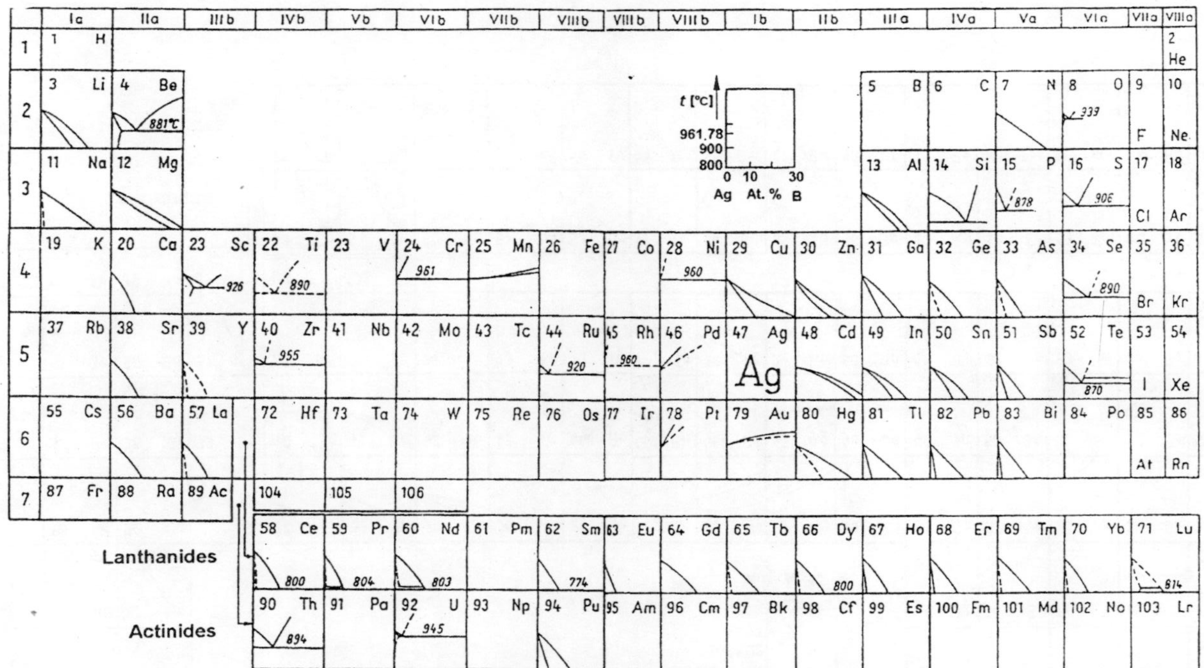

Fig. A5. Types of silver–admixture binary systems in the liquids and solidus regions.

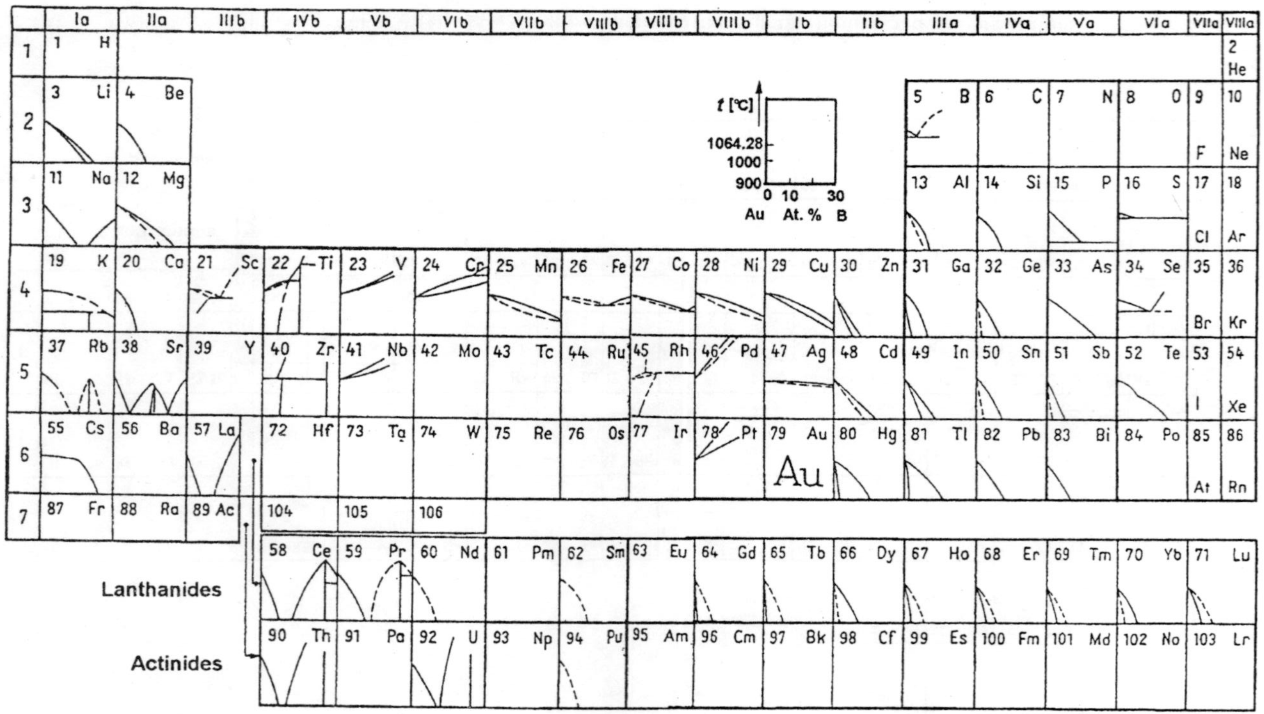

Fig. A6. Types of gold–admixture binary systems in the liquids and solidus regions.

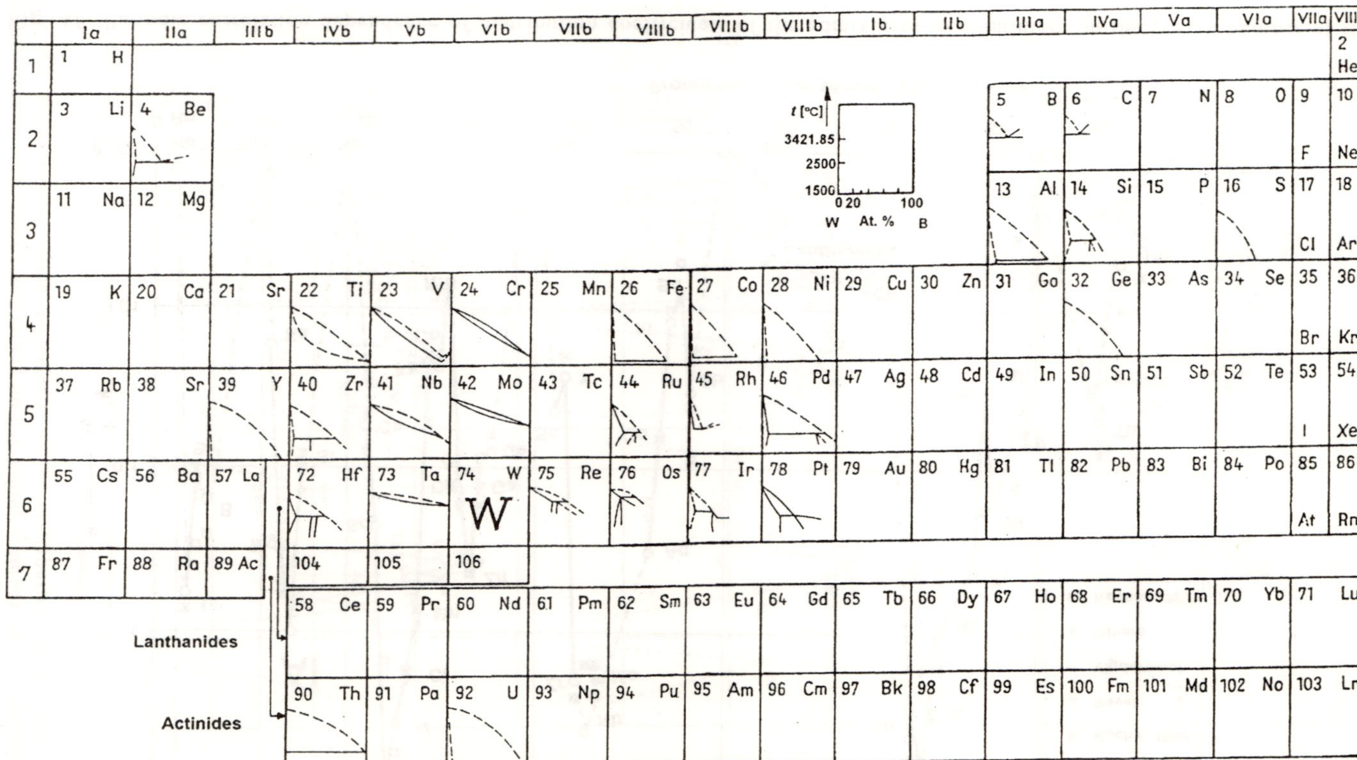

Fig. A7. Types of tungsten–admixture binary systems in the liquids and solidus regions.

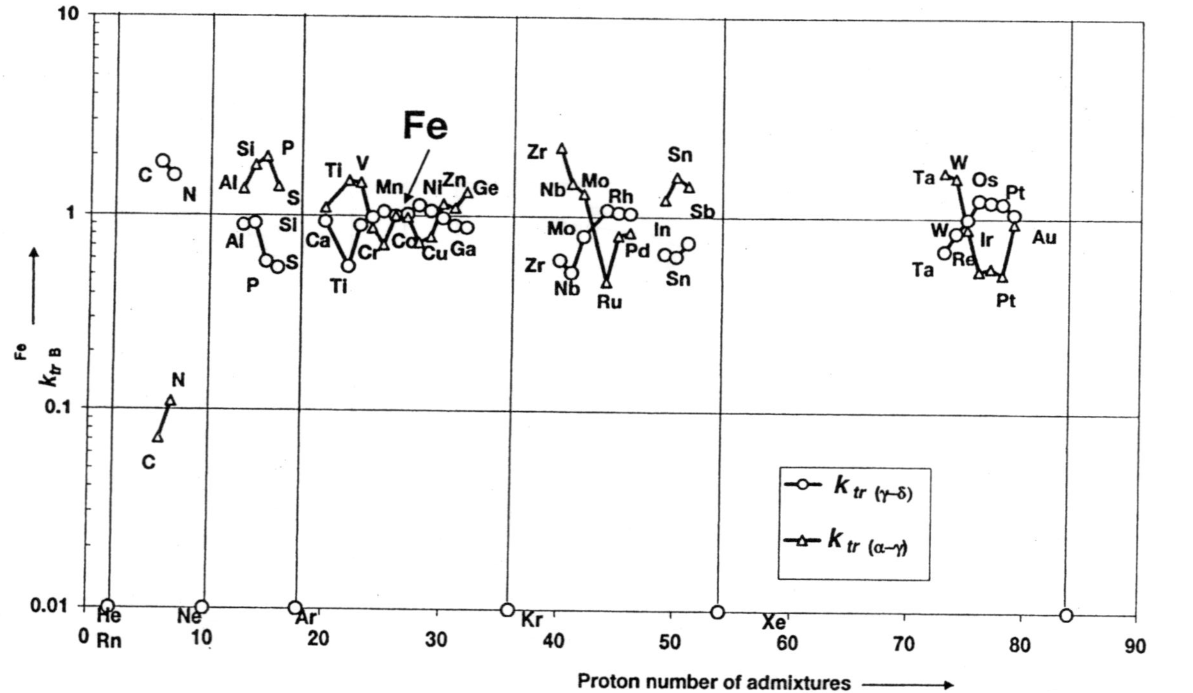

Fig. A10. Periodic correlation dependence of the distribution coefficients of elements in the region of γ–α and δ–γ transformation curves on the proton number of the admixture.

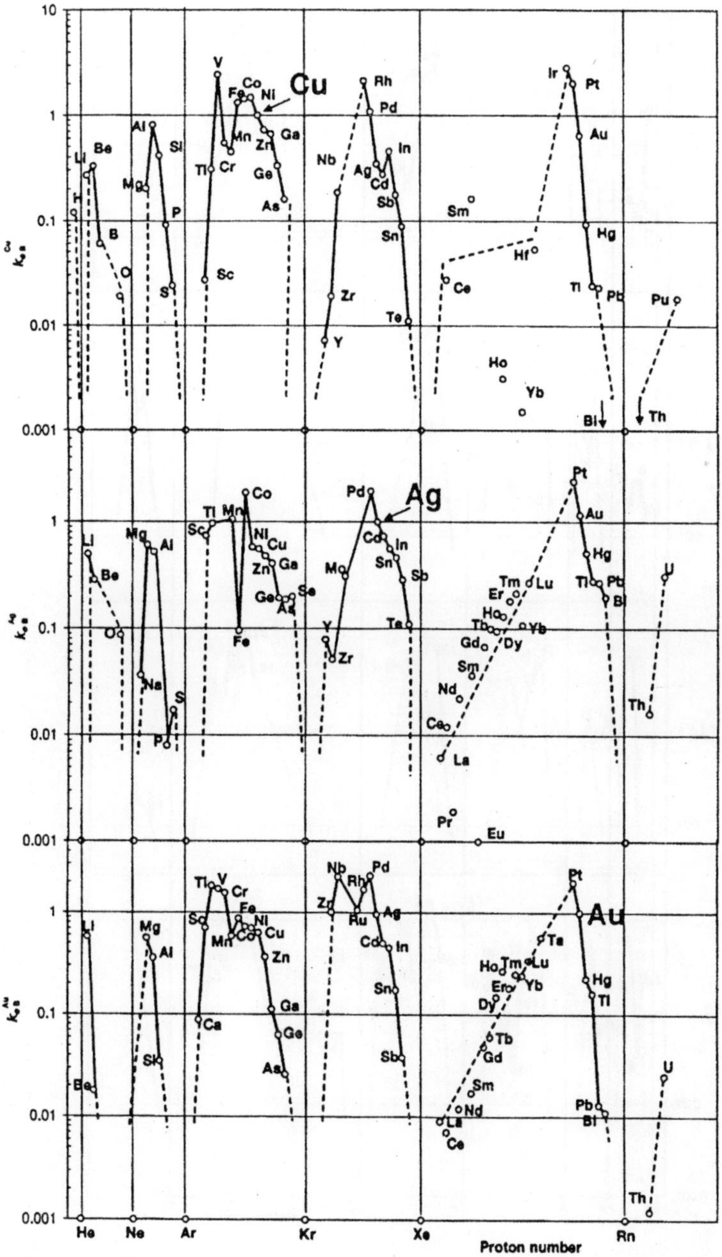

Fig. A11. Periodic correlation dependence of the distribution coefficients of the admixtures in copper, silver and gold on the proton number of the admixture.

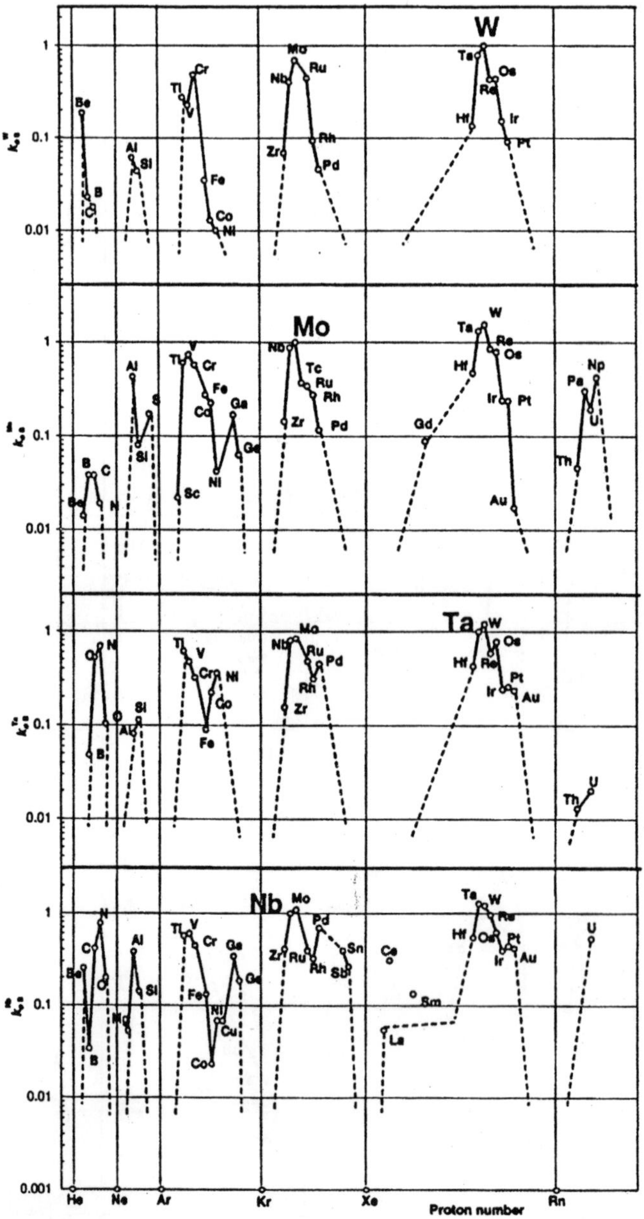

Fig. A12. Periodic correlation dependence of the distribution coefficients of the admixtures in tungsten, molybdenum, tantalum and niobium on the proton number of the admixture.

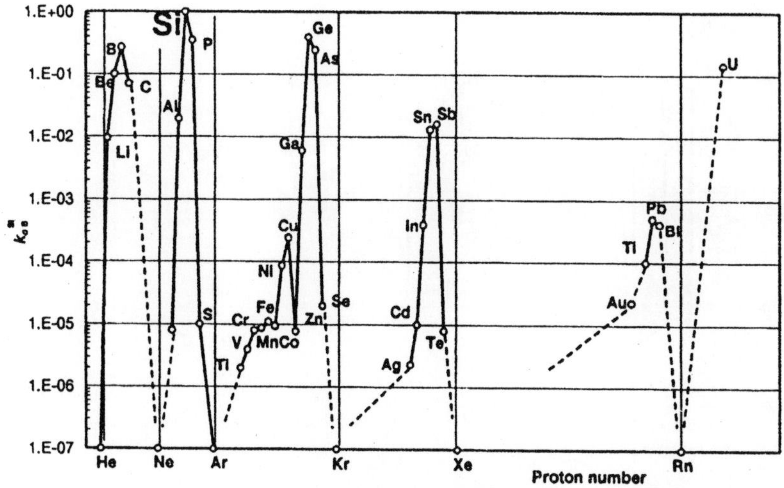

a

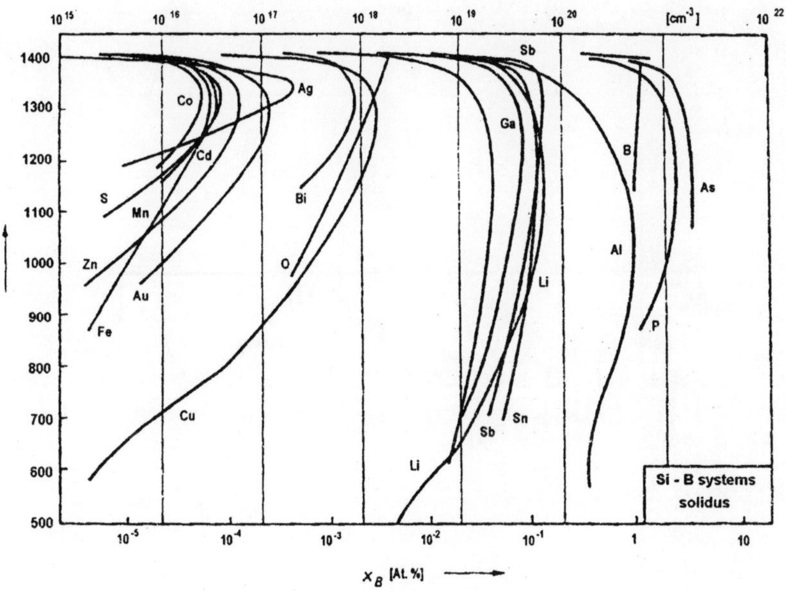

b

Fig. A13. a) Periodic correlation dependence of distribution coefficients of admixtures in silicon on proton number of admixture elements; b) Solidus curves in silicon – admixture binary systems with retrograde solubility.

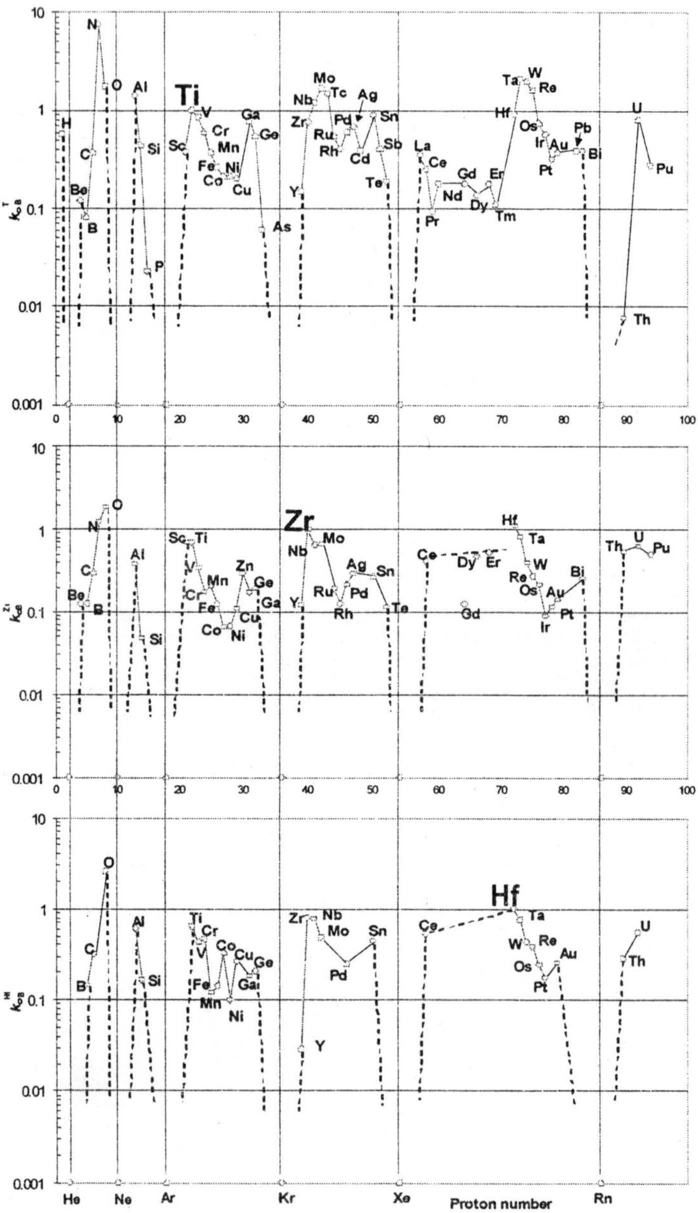

Fig. A14. Periodic correlation dependence of the distribution coefficients of the admixtures in titanium, zirconium and hafnium on the proton number of the admixture.

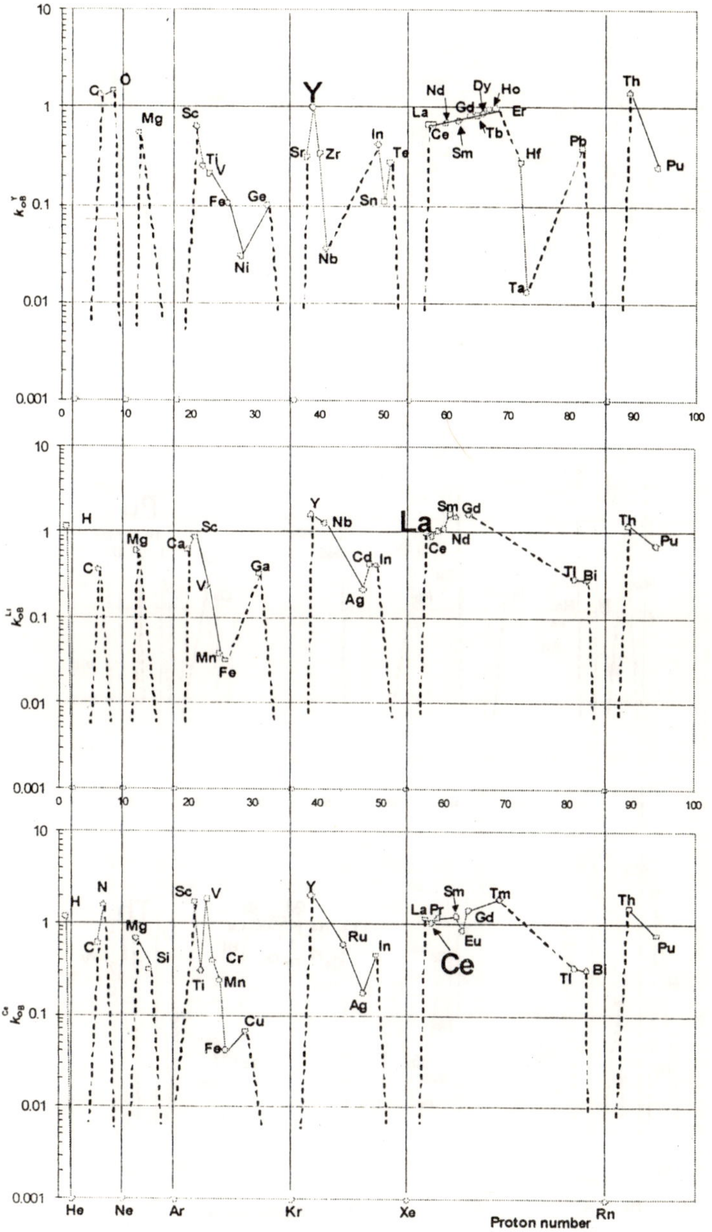

Fig. A15. Periodic correlation dependence of the distribution coefficients of the admixtures in yttrium, lanthanum and cerium on the proton number of the admixture.

Index